The Everyman Project

ROBERT JUNGK

The Everyman Project

Resources for a Humane Future

LIVERIGHT NEW YORK

English translation Copyright © 1976 by Thames & Hudson Ltd.
© 1973 Verlagsgruppe Bertelsmann GmbH/C. Bertelsmann Verlag Munchen,
Gütersloh, Wien First American Edition 1977

Library of Congress Cataloging in Publication Data

Jungk, Robert, 1913–
    The everyman project.

    "Translated from the German *Der Jahrtausendmensch* by
Gabriele Annan and Renate Esslen."
    Includes index.
    1. Technology—Social aspects.   2. Technology and
civilization.   I.   Title.
T14.5.J8613 1977   301.24'3   76–55017
ISBN 0–87140–629–2

Printed in the United States of America

1 2 3 4 5 6 7 8 9 0

Contents

PART ONE

PART TWO: TOOL-KIT

Acknowledgments

My thanks in full would take up many pages, for I am only a link in the worldwide network of friends and co-workers on whose help I was able to count throughout the preparation of this book. I hope they will understand my attempts to save space or to make the material easily understandable, and will not blame me too much for the occasional clumsiness of expression which may have over-simplified difficult and controversial problems.

My especial thanks are due to Professor Eugen Kogon of the Technische Universität Darmstadt, Professor Heinz Haber of Seefeld and Dr Stefan Schwartz of Stockholm, who gave me, as well as advice and information, the encouragement to persevere with this book in spite of my ever-present and still not quite appeased self-doubts.

Finally, I would like to thank the abbot and monks of Seckau Abbey, Steiermark, who, with exemplary tolerance, accepted me in their midst while I was writing this book although my beliefs were other than theirs and let me share something of their peace and their non-materialistic way of life.

PART ONE

Introduction:
At the Turn of the Millennium

The turn of the millennium – flat and colourless words for what used to be called – much more dramatically – the End of the World.

A thousand years ago people were expecting the End of the World. They sold their belongings and prepared for the Day of Judgment whose arrival seemed to be heralded by a continually rising tide of violence.

Modern historians studying the period around AD 1000 have come to the conclusion that it contained the seeds of the enlightenment that began to spread through the following centuries. The limpid intellectual edifice of Thomas Aquinas's thought, the huge shining white structures of the cathedrals thrusting skywards, the teaching of St Francis of Assisi, the spirituality of the humanists, yes, even the critical rationalism of the eighteenth-century Enlightenment – all these things were foreshadowed as early as the tenth century in the work of a few inspired monks hidden away in their cloisters.

The French historian Georges Duby has described the change whose imminence was apparent to only a few: 'Men still worshipped a vengeful magus of a God who ruled and oppressed them. But they were about to refashion him in a more human image, more like themselves, and soon they would be brave enough to look him in the face. Mankind was beginning its long climb towards freedom . . .'

Few people in those days believed that the world could be changed. Their only hope lay in the Kingdom of Heaven. Life on earth was governed by daily suffering and perpetual fear. The Carolingian empire had fallen apart and robber bands roamed through Europe, plundering, torturing and burning all before them. The barbarous climate of the century stifled the seeds of civilization sown in the previous two hundred years, and they perished. Few people were able to read or write.

So the spiritual revival that had begun behind the walls of a few monasteries remained hidden from the contemporary world. Posterity heard of it through the reports of chroniclers like Raoul Glauber, an inconstant, sharp-tongued monk, unpopular with both the higher and the middle clergy. His many enemies called him 'talkative, gullible and inept'. But – so the tradition goes – he regarded these criticisms as praise and as indirect proof that his censures had gone home. Finally he devoted himself entirely to writing down his observations. The monastery of Cluny gave him shelter, and there he composed his five-volume history of the years 900 to 1044.

Chroniclers at the turn of the second millennium are in the same position. Not only should they try to record and criticize the decline and destruction, the brutality and stupidity, the oppression and waste about them, but they should also ask themselves: can we see portents of a coming change? Where could it come from? Shall we survive? I have tried for years to find signs, trends and experiments pointing away from the present as we see it and towards a better future.

At first this was just a sideline which I pursued quite unsystematically: I might come across a newspaper report, a letter, or an oral account of some new possibility or hope, and I would collect such 'good news'. During my years as a correspondent at the United Nations I published privately a few issues of a 'Good News Bulletin'. As a newspaperman, I found it intolerable that the media in their search for news should always report in the greatest detail crimes, catastrophes, crises and wars, while they neglected more positive though possibly less dramatic developments.

These naïve attempts of mine produced a powerful echo from the American public: there were leading articles in major newspapers and news magazines, radio and television interviews, and hundreds of letters from all over the country. They proved how much people longed to hear something other than the daily moans. But my pleasure in this apparent success was short-lived. I soon saw that the superficial interest in 'good news' was no more than a demand for tranquillizers. My readers and correspondents in those days seemed inclined neither to think deeply about alternatives and how to apply them, nor to diagnose the sickness of the age in any profound way. I was drawing attention to a few brighter spots in the prevailing darkness, and they misinterpreted this as an intention to prove that things were not so bad after all.

The extent to which my careful observations were misused for the purpose of covering up existing evils became particularly clear to me when intense interest was shown by those who were especially to blame. I received offers from the chemical industry and airlines to sponsor a daily 'good news' programme on the radio. The broadcaster was to conclude by praising their products and performance!

No, this was not the way to tackle the problem. I saw that I should have to direct my criticism both at those who were preventing any improvement in the existing state of affairs, and at those who allowed them to do so: the string-pullers and the marionettes, the ruthless and the unsuspecting victims. And yet I was haunted by one recurring question: was I really helping to enlighten and mobilize my readers? Or was I, on the contrary, contributing to their sense of resignation? If they came to understand the selfishness, short-sightedness and growing influence of those in power, if they gained insight into the apparently irresistible pressures, into the power structures constantly infringing the liberty of the ordinary citizen – would this knowledge not paralyse them and confirm them in their passive role, so that they would be more than ever content to let everything go on as before? And should I not thus be helping those who know very well how to use others both to cover up and to advertise their activities?

I tried to counter these discouraging thoughts by continually striving to find news about people who were trying to swim against the tide: the Vietnamese, for instance, who drew attention to the suffering of his people by staging a week-long hunger strike on the lawn outside the Palace of Nations in Geneva where the first sitting on Indo-China was taking place. Or Danilo Dolci, the Italian reformer who protested in one of Palermo's worst slums against the Sicilian Mafia and its allies among the Roman politicians. Or the young electrical engineer from Hiroshima, Ishiro Kawamoto, who gave up his career to care for the victims of the atom bomb whom no official body was prepared to help.

But were heroic outsiders such as these really helping to change the ominous course of events? Their sacrifices simply became anecdotes. They only briefly drew public attention to the wrongs that had been committed. Hardly anything was changed thereby. Surely we need to prevent evil before it is done. That is much more urgent than describing or deploring it.

I remember exactly the moment when at last I realized that, as a reporter, I was a kind of profiteer benefiting from the evils of our time.

In the spring of 1960 I was in Japan making a television film based on my book of the previous year, *Children of the Ashes*. We were standing in one of those windy shanty-towns on the edge of Hiroshima. This was where the survivors of the first atom bomb in history had had to seek refuge. We were looking at a couple suffering from radiation sickness. The woman was already so weak that she could not sit up and could hardly speak. The man – white-haired, wrinkled, prematurely aged – had been patiently answering my questions. Now he was asking whether he too might put a question. In a weak voice, not accusingly but more in a tone of apology, he said: 'Did not the honourable scientists of the West realize that their new missiles would go on killing people for decades after they had been launched?'

Since that day I have been obsessed by this Japanese who, fifteen years after 6 August 1945, was dying of the effects of the atom bomb. His question was directed not only at the scientists, but at us all. And it concerned me especially. For years I had been running after events in order to criticize and protest – but always when it was too late. Perhaps I was becoming so dependent professionally on the horrors I described that I was like a doctor whose livelihood is the disease he is supposed to cure. Ought I not to combat my own lack of foresight and vision to try to stop the repetition of a catastrophe like Hiroshima?

Working on a television series entitled 'Europe on the way to AD 2000', I realized that blindness about the future was far more widespread than I had thought. 'I can only think as far as the next budget, or at best the next but one,' an important British politician confessed to me in an interview. 'Anything farther ahead than five years is pretty uninteresting,' I was assured by a man high up in the trade union movement: he emphasized that he was a 'realist', not a dreamer. I met members of parliament who thought only as far as the next election, economists who looked no farther than the next budget, town planners who never planned beyond the next contract.

Very few people in those days understood that economic and scientific developments had set in motion forces whose consequences would stretch far ahead into the future.

And so I began to interest myself in the exploration of the future, and established contact with the first research groups and institutes in France, the US and Japan. At the beginning of the sixties I myself became one of the small band of people who were trying systematically to understand coming events.

Now I began to intensify my search for signs of hopeful future developments. With no special assignment, I travelled unceasingly about the world looking at every kind of research station on both sides of the Iron Curtain – which in those days was still pretty impenetrable. I took part in colloquies and seminars in places as far apart as California and Moscow, Finland and Hawaii, and so became a 'conference hopper' – often regarded with suspicion by the experts, who disapproved of an outsider flitting from meeting to meeting in order to collect the latest papers on research and listen to the exciting conversations that are carried on in the intervals and late into the night. This unofficial part of a congress is usually the most interesting, because it is here that people develop provisional ideas without having to fear their colleagues' criticism. Such ideas may remain mere speculation, but they may also become the accepted facts of to-morrow or the day after.

In the course of these contacts with the avant-garde of science and technology, I kept noticing how little communication there was among the pioneers of the various disciplines. Each man went on delving further and further down his own shaft with hardly any knowledge of what might be going on in the next gallery, let alone the ones beyond that. I rarely found a chemist who cared about sociology, or an organization sociologist concerned about the state of nuclear physics, or a political scientist interested in technological developments – which after all might come to have political consequences. They all said they had more than enough to do to keep abreast of the work in their own special fields. I used to try to tell these specialists about developments in 'other worlds', and make them think about possible connections with their own work; but often they made me feel like an unpopular globe-trotter trying to impress the hard-working worthies of a provincial city with his far-fetched tales of the great world. Of course broadcasts go out night and day, and newspapers, magazines and books are published and distributed in enormous numbers. But the wealth of familiar material overshadows and overwhelms the really new and un-expected, the harbingers of a turning-point: not just the turning-point of the century, but possibly a turning-point in the way we see and understand, in our values, goals and way of life.

I say 'possibly' because it is far from certain that these delicate shoots will grow. They have no chance unless the innovators and experimenters begin to combine forces with the public at large instead of excluding them; unless they let them participate, question and experiment on their own. But such participation will only be-come possible if the present one-way system between experts and non-experts, leaders and led, teachers and taught, the qualified and the unqualified, is opened up in both directions; if more and more people are enabled to take part in deciding, planning and creating, instead of being condemned to the passive role of taking orders and consuming.

In my search for signs of hope, the other important phenomenon was a growing sense of independence and self-confidence among the nameless cohorts of statistics viewed as 'targets', 'socio-economic groups' or 'the electorate' by the powers-that-be. But they too lack contact with one another. There are hundreds of self-help and self-governing projects, of new models for management and schools, of community living projects; but they know little or nothing of one another. Within a single town or country there may be contact between members of experimental social groups, but on the international level there is little communication, comparison, or opportunity to learn from one another's successes or failures. There are not enough generally known examples, be they encouraging or discouraging.

If you look at the spectrum of worldwide attempts to find and develop new ideas and to live by them, you form a more cheerful picture of our future than if you only listen passively to the daily news of disasters. In the end you will arrive at a more optimistic estimate than the computer diagnosticians of the Club of Rome who leave out the two important factors of initiative and imagination because they do not fit into the comprehensible and clearly definable data of their calculations.

In all speculations about the future, man* is the unknown factor. But this should lead not so much to uncertainty as to expectation. While proclaiming the 'limits of growth' – a concept which some commentators compare to the prophecies of the end of the world that were current around AD 1000 – the spiritual fathers of the Club of Rome accompany their study of the future with the words: 'Finally we wish to draw attention to the fact that man must explore himself, his goals and his values, as much as the world that he wishes to change.'

And that is what is beginning to happen all over our planet. We are moving from a period in which the study of nature and the struggle to control it were paramount into a new period where the attempt to understand and develop man himself is becoming in-increasingly intensive. A purely external sign is the rapidly growing number of students in the arts and social sciences, while the number of natural scientists is declining. Every year there is more interest in books about psychology, education, anthropology and sociology. This fact should interest and engage the futurist.

---

* Here and throughout the book I use the word in its root and even strict sense of 'human being', male and female.

There are two chief methods of forecasting future developments. 'Exploratory forecasting' pursues existing trends into the future, whereas 'normative forecasting' uses the insights gained to establish desirable goals and to ask how the gap between the desirable and the possible may be bridged.

The best-known example of normative forecasting was the American 'Project Apollo', which was afterwards severely criticized – probably with justification. This project was planned and announced when the technical apparatus for the moon landings had been only partially invented. But by clearly defining the objective, it became possible to gather together various scattered beginnings and to mobilize constructive forces which were able to bridge the gap by achieving the necessary breakthroughs in the shortest possible time.

Presumably we shall have to employ similar normative procedures when dealing with man's development, for time is running short. True, the idea of growth is already being reoriented to mean inner rather than outer growth; instead of reaching for the world and the sky, people are beginning to look inwards at themselves and their society; but this has not yet become the urgent, generally accepted goal.

One can envisage a worldwide 'Everyman Project'. Its purpose would be to develop hidden, buried or crippled capabilities in countless individuals who have been cheated of their proper development by bad education or social deprivation.

But scientific analysis alone would not be sufficient; what is needed is a large number of plans for new ways for people to live together. Social laboratories should stop being sacred temples where only the high priests of professional expertise are admitted. They ought to be open to spontaneous ideas as well as to patient fundamental research, exposed to continual suggestion, criticism and discussion. They should provide opportunities for experimenting with new types of housing, schools and places for work, recreation and contemplation.

Is this Utopia? Perhaps at first. But what in the past was allowed to remain a dream must now be turned into plans and realities. Alternative life-styles are not just a wish, they are a necessity.

'The turn of the millennium.' These words can be taken as an excuse for retreat, but they can also be a challenge and a task. I have written this book in order to make the general public conscious of the widely scattered, isolated and often hidden beginnings of such a worldwide task. They add up to this: man is not facing his end. Challenged by mortal danger, he is only now reaching his full development.

1 Imagination Saves

Every man a genius ?

In a single night of 1832 the young French mathematician Evariste Galois wrote an intellectual testament so full of amazing new ideas that these few pages, written in feverish haste, have inspired generations of mathematicians. The young man of twenty was facing almost certain death – he had a hopeless duel to fight the next morning – and this enabled him to break through the self-critical doubts and other obstacles which had stood in the path of his exceptional ideas. The great danger he was in might have paralysed another person, but for him it was a liberation that let loose a cataract of productive imagination.

Man on the threshold of the third millennium is in a similarly threatened position. He faces the challenge of life or death. The possible extinction of his species is no longer prophesied in eschatological terms, but by means of curves and statistics. Whatever his colour or his continent, he must face the possibility of catastrophic developments.

But in fact he only plays with such anxieties and secretly hopes to escape. His expectation of a miracle is not focused, as it was a thousand years ago, on a divine power, but on man's inventiveness. 'Some kind of solution will be found' – that is the survival slogan for most of us. But only a very few see themselves as the seekers after effective solutions, or as the inventors of saving possibilities. Imagination is pushed into the role of a *deus ex machina* which will reveal itself only to a few especially gifted men.

The traditional concept of the uniquely creative personality and the chance nature of original ideas is being eroded by experiments and findings which show that every personality contains creative talents. Usually they atrophy for lack of encouragement, acknowledgment or opportunity. But new ideas do not spring only from unpredictable and inexplicable inspiration: there are methods for encouraging them.

If it were possible to open up the untapped reserves of imagination in millions and millions of people, then mankind would find that 'energy for survival' which it sorely needs. In order to awaken these powers, general mass education should lead away from the communication of known facts towards the development of receptivity to that which still remains unknown. Then learning would be more than the acquisition of other people's thoughts and knowledge: it would lead people to solve problems independently, to formulate new concepts and to make their own inventions.

Efforts of this kind can succeed only in an intellectual climate where more trust is shown to the common man than has hitherto been the case. Even many progressive thinkers underestimate the creative possibilities of those innumerable personalities whom they still largely regard as more or less passive objects for instruction. This attitude is not only an expression of supercilious contempt; it also shows irresponsible neglect of an indispensable potential for creative solutions.

The Swiss poet and philosopher Adrien Turel, who invented the profession of 'social physicist' for himself, was one of the first to demand 'everyone's right to be a genius'. He frequently expressed his ideas about the untapped potentialities of the individual in mathematical or physical forms. I shall never forget a night-time conversation in his study on the Limmatquai in Zürich, when he used lightning sketches to show me how most people's lives only move along the 'periphery of the circle' until they have learned to thrust radially into the centre: into the nucleus, where vision and form lie unawakened.

This 'twentieth-century Paracelsus' is not as well known as Fritz Zwicky, another pioneer of the idea that 'everyone is a potential genius'. Zwicky's world renown is based on his work in astrophysics at the Mount Palomar Observatory in California, and especially on his discovery of exploding stars (supernovae) – a discovery which sprang from a completely new method of observation. His 'morphological method', which underlies many of his pioneering discoveries, is less familiar; but Zwicky claims that it has led him to systematize the process of invention.

The method begins with the exact description and generalization of a given problem. It then defines the elements of the problem and proceeds to arrange them in unpreconceived combinations; and it led Zwicky – a Swiss working in America during the war – to make one of the most important discoveries in the field of rocket propulsion. He made this discovery by systematically examining 'every possible form of chemically driven jet propulsion which can maintain or even accelerate its movement through a vacuum, through the atmosphere, through water or through the earth'. Zwicky investigated 576 possible types of jet propulsion, including theoretical inventions like the 'terra machine' which have not yet been put into practice. He maintains that the latter will play the same part in exploring the earth's interior as rockets have done in exploring the sky.

Zwicky proclaims: 'Every morphologist is a professional genius... who can make discoveries in every field of science, technology and life.' He goes even further: 'Few people today recognize the fundamental fact that everyone is a potential genius and that a sane, free and healthy world can only exist if all these geniuses are recognized and allowed to develop.'

However, modern creativity research does not fully confirm such wide-ranging assertions, made under the continuing influence of the nineteenth-century cult of genius. But even these more sober experiments do show that the faculty of inventing ideas is much more widespread than had been supposed. But we do not encourage the gift of original thought independent of norms and routine, present to a large degree in children; on the contrary, we teach them to adapt to the 'rules' of the adult world, and they lose their aptitude.

Educational researchers like the American Paul Torrance, the Englishman Edward de Bono, the German Günther Wollenschläger and Erika Landau of Tel Aviv have given us glimpses, through their experiments, of this paradise of the imagination which is lost by each generation in turn; they have asked themselves the important question of how these faculties can be preserved and can help to solve the problems of social reality. Their attitude turns the process of socializing the personality into a creative rather than an imitative one: each person begins in his own way and is taught to give rather than to receive.

The English historian Arnold Toynbee recognized how important this change of direction could be for the future of mankind. He writes that it is a life-and-death matter for every society to provide sufficient opportunities for the development of creative abilities. We lose billions of these opportunities. But all the same, more and more schools and teachers are choosing the new path: young people are encouraged to observe independently, to judge critically and to take inventive risks: they are being prepared not for the existing order, which demands conformity and obedience, but for a society which is only just emerging.

For centuries the rising flood of imagination has been confined chiefly to the arts and technology; directing it towards social goals will demand even greater and more strenuous efforts than does the creation of individual works of art or pieces of machinery. In order to survive, man at the millennium will have to 'invent' more humane ways of living, more open institutions, more flexible modes of behaviour. He will change the existing course of progress and aim for other objectives, because the present direction is turning out to be increasingly dangerous. The example of those who gave early warnings and suggested new orientations when the majority were still blindly pursuing 'limitless progress' cannot be decisive, but it can help.

There was one man who came close to fulfilling my idea of a person thoughtful and decisive enough to master the crises at the turn of the millennium. This man was a mixture of sagacity and imagination, fear and courage, despair and hope. He understood what is approaching us. He himself had helped to set it in motion; later he wanted to stop it, although he often despaired of that being possible. His motto was this: 'On paper an ultimate catastrophe brought about by man seems 85 per cent likely. I live for the other 15 per cent.' The man I mean was Leo Szilard, who died in 1964. He was a chemist, physicist, cyberneticist and biologist, and according to his collaborator John R. Platt his social and political influence was such that 'in a hundred years he will appear to have been the most influential personality of our time'.

How can one explain this high opinion, which is shared by other historians of science? At first it must seem a flattering exaggeration, for the name Szilard is not widely known to the general public. This is not surprising, for he was a man who worked behind the scenes. He contented himself with the role of sower, the supreme stimulator of ideas who would leave the working out of his conceptions to others, and the fame and wealth that went with them as well. Szilard's inventive genius was fruitful not only in his own field of physics, but also in other scientific disciplines such as information theory, biology and genetics. And beyond that lawyers, economists, strategists and particularly international politicians are indebted to him for fundamentally new ways of looking at things.

Szilard's pioneer role in bridging the gap between science and society was at least as important as his contributions in various fields of research. He was among the earliest scientists to realize the growing influence of science upon the course of history. In 1933 the discovery of the neutron convinced him that the atom would soon be split, and long before the war he persuaded English and American atomic scientists to keep their work secret. He did not want Hitler to be the first to have that devastating weapon, the atom bomb, whose construction he foresaw. That is why, in 1940, together with his fellow-Hungarians Eugene Wigner and Edward Teller and the Italian Enrico Fermi, he persuaded Einstein to write the fateful letter that decided Roosevelt to start what was later called the Manhattan project.

But when the bomb was ready, Szilard again saw farther than his colleagues and warned against its military use as a weapon for mass destruction. He had urged America to build the bomb only as a deterrent, so that no one should ever use it. If Hitler's scientists should succeed in making such a super-weapon, he would never dare to use it for fear of reprisals. But this stalemate never came about. The US Air Force, in sole possession of the monster, bombed Hiroshima and Nagasaki.

Szilard now began a long campaign to get the genie back into the bottle. He became the most important initiator of the movement to withdraw control of atomic energy from the American military, and in 1952 he helped to found the East–West Pugwash Conferences which laid the foundation for an understanding between the US and the USSR at the coldest moment of the cold war. In 1961 he founded the Council for a Livable World, one of the first civil rights movements, which demanded that the man in the street should have a voice in the further development of the industrial revolution and its political consequences.

This millennial man Szilard never had a fixed abode. He lived in hotels or in hostels for scientists like the Quadrangle Club at Chicago University, where I first met him. He never accepted a permanent appointment; he played each laboratory as a guest artist, collected collaborators, inspired them, and then moved on. It was said that in his search for new ideas he spent less time at his desk or in experiments than in the bath, where he found regular inspiration. Wherever he was, this volcano of words and ideas drew everyone into conversation. In Washington he based himself in the hall of a modest hotel on Dupont Circle. Anyone could come in from the street and talk to him.

He liked trying out his ideas on people whom he barely knew. If they only nodded and agreed, then he suspected that something must be wrong. Probably his idea had been 'too normal'. He proclaimed that new concepts were rarely logical extensions of accepted ideas: they sprang from the subconscious and should therefore present a challenge to what was generally held to be reasonable.

These improvised debates led to many important moves, such as the demand for a twenty-fold increase in American research on birth control. Szilard was one of the first to grasp fully the problem of the population explosion. He also initiated the idea of creating special scientific task forces not just for one but for all of the crises facing mankind. After his death this task was carried on chiefly by his former student John R. Platt.

Szilard hoped that mankind would be saved by scientists working together across national and ideological boundaries. His belief in the world-shaking power of this statistically minute sector of the population had been confirmed by the drama of the atom bomb. But was he not overestimating the influence of the researchers? And further: was there not an anti-democratic concept behind these ideas which would make a small number of grey eminences and the powerful men they advised responsible for decisions which concerned everyone?

Here we see the beginnings of a conflict which is becoming ever more clear in the critical years leading up to the millennium: should – or must – the blueprints for the future be drawn up by the social imagination of a small, highly qualified élite? Or can social imagination be democratized and made to rest on a wide – the widest possible – base?

Discoveries we must make

At the beginning of 1969 I found an airmail letter from the university town of Ann Arbor, Michigan, in my mail. Its contents were so exciting that for days I was hardly able to sleep. It contained a duplicated copy of a paper by John R. Platt entitled 'What We Must Do'. It was an invitation to every scientist concerned about the fate of the world to meet other researchers in 'working groups for crisis studies' in order to develop a strategy for research to counter the dangers threatening man's survival.

According to Platt, not just one but many simultaneous crises loomed over the millennium: crises in armament, population, participation, environment, race, famine, cities, energy. But the decision-makers all over the world were unable to see which defensive measures were urgent and which were not; neither had they sufficient ideas and concepts to deal with this unique historical situation. The situation was just as acute and serious as during the

Second World War, when all scientists were mobilized, with the result that this concentration of exceptional minds produced decisive breakthroughs in the field of arms technology.

Platt said:

> I believe we are going to need large numbers of scientists forming something like research teams or task forces for social research and development. We need full-time interdisciplinary teams combining men of different specialties, natural scientists, social scientists, doctors, engineers, teachers, lawyers, and many other trained and inventive minds, who can put together our stores of knowledge and powerful new ideas into improved technical methods, organizational designs, or 'social inventions' that have a chance of being adopted soon enough and widely enough to be effective. Even a great mobilization of scientists may not be enough. There is no guarantee that these problems can be solved, or solved in time, no matter what we do. But for problems of this scale and urgency, this kind of focusing of our brains and knowledge may be the only chance we have.
>
> Scientists, of course, are not the only ones who can make contributions. Millions of citizens, business and labor leaders, city and government officials, and workers in existing agencies, are already doing all they can to solve these problems. No scientific innovation will be effective without extensive advice and help from all these groups. . . . But the step that will probably be required in a short time is the creation of whole new centers, perhaps comparable to Los Alamos or the RAND Corporation, where interdisciplinary groups can be assembled to work full-time on solutions to these crisis problems. Many different kinds of centers will eventually be necessary, including research centers, development centers, training centers, and even production centers for new sociotechnical inventions. The problems of our time – the $100-billion food problem or the $100-billion arms control problem – are no smaller than World War II in scale and importance, and it would be absurd to think that a few academic research teams or a few agency laboratories could do the job.

I had met Platt a few years before his appointment to the Mental Health Institute at the University of Michigan. At that time he already spoke of the need to develop a 'science of survival'. Now he was seeking to convert his ideas into practice. The Institute where he worked is not – as the name suggests – devoted to research into mental illness, but seeks rather to investigate the social preconditions of mental health in our epoch of swift change often beyond intellectual control. Thus it was entirely in line with the founders' views when Platt began, from this base, to start 'crisis study groups' in many universities. At least once a month, newsletters full of questions and suggestions were sent out to the other groups, including for example:

What new 'social indicators' do we need, as for monitoring 'humanization' of administration, or 'crisis-handling effectiveness'? (Good indicators are an enormous help in solving problems, because they give 'error-signals' that permit maximization.)

Could we set up any kind of reward-system for social inventions – the counterpart of the 'patent system' for technological inventions – so as to maximize individual incentive to make such inventions.

Could we set up an additional parameter of control in the economic system (besides Federal Reserve and tax rates) that would decouple prosperity and full employment from inflation? Inflation squeezes the poor and the government servants, multiplies labor–management wage crises, and creates widespread cynicism and despair over 'progress'.

Design of a peace-keeping international structure, using new ideas of feed-back stabilization and self-reinforcing loops and lock-ins, to give both more acceptability and more effectiveness and sense of security.

Other ideas which were discussed and formulated by these groups included new methods for teaching, for child-rearing, for decentralizing management, for non-authoritarian types of organization, for freeing women from housework, for a more general use of ombudsmen, and for greater general participation in politics and the economy.

Three years after the publication of 'What We Must Do', in August 1972, another article by Platt, this time in collaboration with Richard A. Cellarius, appeared in the weekly *Science* (Washington). It appealed to the world at large, and proposed to set up 'councils for urgent studies' in as many countries as possible, based on the experience of the first crisis study groups. Their work was to be co-ordinated by a sort of permanent 'general staff for mankind'.

An international survey of this kind would be able to pick out dangerous gaps in the general struggle for survival. Not only the specialists, but the world public, should be made aware of weak spots.

Platt, Cellarius and their colleagues have already produced a 'map of the most urgent research priorities'. The 'front' is divided into twenty-five sections where essential discoveries and inventions are needed. Some of these, says this provisional 'general staff', have already been 'discovered' by the world public and have become – albeit rather late – the subject of the necessary intensified activity. They include, for instance, the discovery of new sources of energy, improved utilization of or invention of substitutes for important raw materials, control of the population explosion, and the struggle against pollution. But at least a third of the most urgent areas for research are still undeveloped or insufficiently recognized. These include organizational science, mass communications, the study of social change, the examination of 'big systems', and peace and futurology research.

In view of man's precarious position, intense research is proposed in the following areas:

◇ the encouragement of social inventions;
◇ forecasting;
◇ generally intelligible information about critical situations;
◇ popular participation in decision-making;
◇ disaster research;
◇ methods for peace-making and peace-keeping.

Platt sees a network of working groups for urgent studies based on the universities of the world. The next step would be national groups consisting of members of the public as well as academics. Finally the 'general staff' should become an international institution working with the United Nations; it should function as an early-warning system for the whole planet and maintain a continual watch on the future. It should ask from what direction the greatest danger threatens, and where research is most urgently needed to come to the rescue.

Certainly mankind has hitherto lacked a worldwide political and social weather station of this kind. The UN's efforts to create one in 1971 never got off the ground, first of all because the necessary funds were lacking and secondly because of the not totally unfounded suspicion that this might lead to a sort of scientists' dictatorship which would consider itself the sole repository of reason and, conscious of its own superiority, would try to manipulate the nations as well as their statesmen.

Here and there Platt's ideas have fallen on fertile ground, even if not to the extent that he hoped. The International Institute for Applied Systems Analysis at Laxenburg, near Vienna, where researchers from East and West mull over the crisis problems of the world, could be a useful step in the right direction, albeit not wholly without danger for the freedom of political decisions. The study groups that have developed from the Club of Rome founded by the Italian industrialist Aurelio Peccei point in the same direction. Slower progress is being made by the world trade union movement in setting up institutes for worldwide surveys and forecasting in the interest of its members in order to counterbalance the rule of experts and the technocracy. The multinational corporations associated with Herman Kahn and his Hudson Institute were much quicker off the mark. They have sponsored a study on 'The Prospects of Mankind' which looks far into the twenty-first century. As far as one can tell from the preliminary studies, it will be an unabashed defence of further industrial growth and the monopoly capitalism of the multi-national industrial giants.

Of all the suggestions published by John Platt in his manifesto-like studies, the most fruitful is his plea for 'social inventions'. What he means by the expression, and what he considers the most important innovations in the field of human and social studies between 1900 and 1965, are specified in a study he made in collaboration with Karl Deutsch of Harvard and Dieter Senghaas of Frankfurt University. The criteria for their choice are given as 'new conceptions of relationships, leading to new means of behaviour – that is, they must have helped people to perceive something which they had not recognized before, and thus led to new discoveries. . . . Or they must have made it possible to do something which had not been done before.'

The sixty-three selected 'major advances' in social innovation include thirteen new concepts in psychology, twelve in economics, eleven in politics, eleven in mathematical statistics, seven in sociology, six in philosophy and three in anthropology. The different topics include psychoanalysis and depth psychology (Freud, Jung, Adler); intelligence tests (Binet, Terman, Spearman); non-violent political action (Ghandi); play theory (v. Neumann, Morgenstern); guerrilla organization and government (Mao); the study of small groups (Lewin, Lippit, Likert, Cartwright); general systems analysis (Bertalanffy, Rashevsky); opinion polls (Gallup, Cantril, Lazarsfeld, Campbell); computer simulation of social and political systems (McPhee, Simon, Newell, Pool, Abelson); cost effectiveness calculation (Hitch); conflict theory and multisum games (Rapoport).

The concern here is with absolutely fundamental innovations. That is why the study does not deal with less important social inventions such as credit cards, think tanks, the Common Market, assembly lines, shopping centres, collective farms and kibbutzim.

Platt and his two collaborators undertook their study chiefly because they wanted to discover what conditions favour the conception of new ideas for the further development of man and society. They found that universities with wide opportunities for the exchange of ideas among members of different disciplines were the most fertile. They examined the age of social inventors and found that 160 of them had been between thirty-five and thirty-nine years old at the time of their decisive research. More than 40 per cent were between forty and forty-nine, and only 6 per cent were over fifty. They also went into the question of individual research versus team work, and found that almost two-thirds of the important new initiatives were made by individuals. But their role is declining: before 1930 three-quarters of all contributions were made by individuals, after that less than half. The conclusion reached is that in the next decade teams of social scientists will probably be the main initiators of progress.

It is strange that a study which examines in such detail the conditions favourable for creative imagination should not consider why there are so many fewer new ideas and concepts in social science than in natural science.

Platt himself mentions one of the reasons in another context: there are no material rewards for social innovations, no patent laws, and hardly any commissions. The inventors of social insurance, permanent education, shift work, minimal wages and the hot line between rival great powers have probably done more for the improvement and greater safety of our lives than many technical inventors; but they received no material rewards, and only a few exceptions received recognition. And there are other obstacles more difficult to overcome.

In September 1971 I attended a large seminar in Rensselaersville, upstate New York, which had been organized by UNITAR (United Nations Institute for Training and Research), one of the smaller auxiliary organizations of the UN. Its purpose was to consider setting up a UN authority for the continuing consideration of the long-term problems of mankind. One of the themes we discussed with a view to future developments was aggression and individual terrorism. What innovations, what aggression substitutes could be found? Would it be possible to solve the problem of political force by social innovations (world games, for instance, in place of world crises and world wars)? We soon saw that the opposing cultural and ideological attitudes of the participants were bound to prevent us from making common alternative proposals.

Technical innovations are quite wrongly considered neutral, whereas alternative experiments in the social field are restricted from the outset by political and ideological commitments. Beyond that there is the question of fashion and the spirit of the age. Even innovators are confined within the prison of their times, from which they can escape only with difficulty, if at all. The heretics and reformers at the end of the Middle Ages, for instance, could only imagine a world with a different Church, not a world with scarcely any Church at all as we see it today. Similarly, the people of our era can imagine an age with a different kind of large-scale industrial enterprise, but not with none or almost none. Even the apparently radical ideas of revolutionaries are largely influenced by the milieu against which they rebel. And yet their refusal to adapt, their 'no' to the existing order, influenced by that order as it is, represents the first step – but only the first – towards completely new possibilities. The first condition for a prison break from existing rules and accepted norms would be the emphatic encouragement and promotion of 'deviant' thinking. That was the promise of creativity research. What has happened to it?

The creativity boom

Productive Thinking (New York and London, 1945), a book by the German psychologist Max Wertheimer who emigrated to the US from Nazi Germany, showed that the creative spirit is not the privilege of a specially chosen few but can be developed through practice. The tensions set up by a problem can become the impulse towards new concepts for whoever tries intensively to tackle it. According to Wertheimer, productive thinking begins with the analysis of a conflict, with the question: 'Why won't it work?' When the problem has been defined, the next step is to try and solve it by available or unavailable means. The question 'What have I got at my disposal?' leads to the analysis of the ultimate purpose, an analysis which asks: 'What do I need? What can I do without?' Anyone going through these steps creates the conditions for new judgments, for improvements, possibly even for genuine inventions.

This idea was received with interest by American specialists in the field, but it was not for another five years that the chief impulse for studying human creativity came from a single lecture given by

Professor John P. Guilford when he was president of the American Psychological Association in 1950. It was he who first introduced the word creativity to the general public.

Guilford showed that independent and unusual thinkers scored badly in the usual intelligence tests because instead of writing down or ticking off what was expected they followed their own ideas. This meant that the most creative people slipped through the nets of the tests.

Guilford's plea to pay more attention to the neglected problem of creativity and to study it in depth found a strong echo. In the US alone it was followed by a flood of work on human creativity ranging from the excellent through the mediocre to the downright dubious. By 1955 there were fifty-three publications a year, in 1965 the number had risen to 474, and in the eighteen months from January 1965 to June 1966 a bibliography issued by the newly founded Creative Education Foundation listed almost three times as many again.

Interest is not confined to the US: in the Val de Chevreuse, near Paris, the French National Research Council has assembled a team of artists, teachers and managers to discuss the phenomena of creativity. In Kyoto the planners of a leading electronics firm have been told to 'invent future historical epochs'. The new vogue produces previously unthinkable and sometimes unwittingly comic situations: in a large London firm the leader of an intensive session told his serious, somewhat stuffy group: 'Each of you is a virus. Yes, part of a virus culture making itself at home in a part of the human body. How does it feel?' Even in Moscow, creativity has become the subject of the day: students of the Institute for General and Technical Education have to watch a coin lying on the edge of a spinning disc. 'How can you stop it falling off?' asks their professor. In Poland the first conference on creativity was called in the fifties, and Professor Alexander Matejko asked the highly important question of whether social and environmental factors might not be more important for the development of creative talent than the elimination of individual psychological obstacles which preoccupied American researchers.

Guilford himself was surprised by the overwhelming response to his plea that 'divergent thinkers' should be picked out and their talents fostered. The response was due to the mood prevailing in America at the beginning of the fifties. A terrible war had ended, but hopes of peace and real improvements in the human condition had been disappointed. In this atmosphere of disillusion, scepticism and political paralysis, the promise of new powers to be discovered and developed appeared as a hopeful goal – a substitute, perhaps, for the social change many people longed for.

But Guilford was already working in a new direction: he was thinking of using the discovery and development of creative personalities mainly as a means of improving America's position in the cold war.

We must remember the situation at that time: the US had been in a dominant world position above all because of its monopoly of the atom bomb. This position was neutralized by the first successful Soviet atomic weapon trial in 1949. A stalemate situation developed, in which the Americans tried everything to regain the initiative by means of breakthroughs in science, technology or international politics. It was the beginning of the era of think tanks, those extraordinary set-ups where hundreds of brilliant minds from all disciplines were hired to analyse problems and come up with something new in the interests of American defence. This faith in intellectual activity supported by ample funds was something new in American life. The military, of all people, with their mistrust of self-willed, undisciplined eggheads and long hair, were admitting these people into their closed society. This move can be explained by their recent experience: it was scientific intelligence, with its technological and organizational discoveries, that had won the Second World War for them.

Compared with the huge think factories like RAND and SRI (Stanford Research Institute) with thousands of research workers, Guilford's Aptitude Research Project at the University of Southern California in Los Angeles was only a small think workshop. It was commissioned by the Navy to test potential officers for their aptitude. The work convinced Guilford that the usual evaluation by intelligence quotients was unsatisfactory. Nine years after the lecture which by now had become known as 'the historic trigger of the creativity movement', Guilford asked himself why he had had such a powerful effect. He thought the military motive was the most important factor – the fact that the US was in the throes of a deadly struggle to safeguard its way of life. The military side of this struggle, he said, within the framework of the world competition for new weapons and new strategies, demanded a higher rate of invention.

In order to forge and use their new weapon 'imagination', the Americans, inspired by Guilford, immediately financed hundreds of projects to discover the divine spark. And when in 1957 not they but the Russians, with their Sputnik, achieved a breakthrough in the technology race, the Americans redoubled their efforts.

So the imagination was unfettered only to be immediately chained up again and put into uniform. Its objectives were to be new nuclear warheads, plastic bullets invisible to X-rays, electronic battle robots, chemicals for defoliating the jungle, and so-called harmless gases to be used against demonstrators. The production target was a reliable annual output of Edisons, Steinmetzes and Oppenheimers.

Did it succeed? A retrospective study on the use of technical innovations (Project Hindsight) seems to show that really new ideas did not particularly flourish in this milieu. Imagination was adapting to the unimaginative.

More about the effects of putting creativity to work can be learned from its use in the battle of industrial competition. Three years after Guilford's starting signal, the creativity boom in this sphere was initiated by a work called *Applied Imagination* (New York, 1953). The author, Alex F. Osborn, was a member of a leading American advertising agency. He promised to teach his readers the principles and procedures of creative problem-solving. Osborn advocated a new group method for developing the imagination which he called brainstorming. A year later his book was in its fourth printing, and by 1965 there had been twenty. Hundreds of firms arranged brainstorms for their employees. The key phase of this process occurs when the participants are told to switch off their critical judgment and encourage each other to produce as many crazy ideas as possible. In a later phase these ideas are examined for any possible utility they may have.

The later editions of this book are full of triumphant accounts of victories won by the new method:

◇ Dr Fred C. Finsterbach organized a session for American Cyanamid in which 92 ideas were produced in 15 minutes.

◇ General Electric tackled the problem of how to get more suggestions for improvements from its employees. In 35 minutes they had 131 ideas on this subject.

◇ The television division of the Radio Corporation of America got 200 suggestions for improvements in a single sitting for its viewers.

The method quickly spread overseas:

◇ A course for store managers in Melbourne, Australia, produced 163 suggestions in 40 minutes. Sixteen of them were immediately applicable.

◇ A life insurance company in Berne, Switzerland, asked: how do we recruit more suitable representatives? In 60 minutes there were

225 replies; 125 of them were partly usable, 25 were used immediately.

◇ The world-famous Swiss chocolate manufacturers Lindt–Sprüngli organized a brainstorm to determine what kind of foundation they should set up to celebrate their 125th anniversary in a way that would have the maximum public effect. In a short time nearly 400 suggestions were collected.

But the sociopsychological side-effects of brainstorming are at least as important as the flood of new ideas. Ossified authoritarian structures within the firms are broken up – at least temporarily – by playing with the imagination; inhibitions of many years' standing melt away. The creative group feels more strongly than any single employee working on his own that everyone is doing something jointly, something unusual, but enjoyable and possibly even useful.

However, a detailed analysis showed that even in experiments such as these, rivalry and competition hinder the flow of creativity and frequently cause it to dry up. Using videotape recordings of ideas sessions, George M. Prince was able to pinpoint the inhibitions set up by these feelings.

H. Geschka and H. Schlicksupp of the Frankfurt branch of the American Battelle Institute (Research Group on Creativity and Imagination) spent two years collaborating with seventy firms on methods of idea-finding. They discovered thirty-two new methods, and a further twelve through more than two hundred experiments. They sum up as follows: 'The methods for idea-finding can be roughly divided into two groups. . . . The first group is analytical-systematic (e.g. morphological boxes, progressive abstraction); the second stimulates the imagination (e.g. brainstorming, synectics).' One of the most important conclusions of these significant studies is that different problems require different idea-finding methods.

According to this survey, the most original ideas are developed by the technique of 'synectics', which was invented by the American industrial psychologists William J. J. Gordon and George M. Prince. The idea is to stimulate by conversation processes of thought and association which normally only emerge from the subconscious in dreams. Five to seven people form a group and try to think themselves into a strange and bizarre situation by using unusual combinations and analogies in their talk. An 'imaginative analogy' might throw out the idea: 'Could we construct a new ready reckoner by employing trained ants?' This crazy question provides opportunities for overcoming mental obstacles and clichés behind which completely new and possibly feasible ideas may be hidden.

I visited the Synectics group headquarters in a two-storey house in Church Street, Cambridge, Massachusetts – a quiet side street that seemed more English than American. I harboured a prejudice which turned out to be unfounded. In the neighbouring academic milieu, I had constantly heard two objections levelled at this research unit: firstly, that the work there was 'unscientific'; and secondly, that the group had been corrupted by commissions from industry.

Both these objections are based on undeniable fact, but do not take account of the causes. Gordon and his colleagues (who originally included the university professors Donald Gifford, Richard Foster and Jerome Bruner) began by observing the creative process at its source in order to discover the psychological mechanisms behind it. Their purely scientific efforts were gradually replaced by more and more assignments from industry. Without them, the unit could not have continued its work.

Originally they had expected financial support from the American universities. But these had been unable to stomach Gordon's unconventional methods, and left him to find a less prejudiced patron in industry. As early as 1944, he had begun to experiment on research workers and inventors who were under analysis. They were told to admit the free flow of memories and associations which they had learnt to use on the couch into their efforts to solve technical problems. This led to unique monologues in which the moments of creative inspiration and the detours that had led to them could be pinpointed and interpreted by means of comparisons.

A researcher was working on an improved altimeter for aircraft. It was to be easy to read and not liable to loss of precision through wear and tear. This is his inner monologue:

> . . . I am taking apart this airplane altimeter. Why! There must be over a hundred little gears in this mess. . . . I notice that springs are essential. I can imagine throwing out any other element but this master spring . . . this foundation spring. Any isolated unit for measuring altitude by pressure will have to be backed up somehow by one or more springs. . . . Thinking about springs, I find that I feel very much removed . . . sort of cut off from the unit I'm playing with, even though it's right in front of me. . . . It seems to be disintegrating almost of its own accord. There are the pieces . . . what piece can I least afford to throw out? The spring is central here . . . but what is the spring? What does it mean to say of a spring that it is a spring? How would I feel if I were a spring? I find myself very mixed up with this spring. I

can't get away from my own springiness . . . even if I wanted to. But I don't want to. I am folding in and then expanding, folding in and then expanding . . . or I'm being pulled out and then I'm being pushed back in. Tight! This I don't like. I don't like this tight feeling. . . . I don't like this feeling of being pulled way out either. Someone's got me by the hands and the legs, stretching me over a rack, torturing me. . . .

What if an altimeter just were a spring? . . . No, that couldn't be. . . . You can teach a spring to do something by building in some kind of response . . . even more perfectly than you can train a child . . . except for hysteresis. I wonder what hysteresis would be in a child? I'm very sorry, Madam; your child has advanced hysteresis and we've got to operate immediately. . . . How do you get inside that spring? If I . . . if there were an enormous spring . . . a spring as big as a house, and I hold onto it and it goes in and out, in and out. What happens to me? Well, let's see, I can put a little drop of ink on the thing. Now I wind it in and out. Look. If I get a spring big enough, as the spring tightens . . . the blob of ink will move in . . . and out as the spring relaxes. . . .

Funny . . . now I have the feeling that this thing is on its own, completely outside me . . . that the whole idea is no longer . . . no longer anything to do with being mine, my idea . . . it's just like flying now because if I put a spot on the spring and tighten it up, it performs an arc which is exactly the band I'm looking for. It's amazing, and this is no longer I . . . it's as though it was taken away from me. . . . This must be what people mean when they say you start writing a play and the people you've put in the play just go on by themselves.

Similar experiments were made with artists who gave free reign to their thoughts, fantasies and ideas while they were at work. The experiments showed parallels between scientific and artistic creative processes. Later this work led to the so-called 'rock pool experiment', in which 'thinking out loud' was first used in group conversations. New methods were discovered for forcing the creative process. The main technique is to make the familiar strange, and the strange familiar, a process as simple as it has been successful.

I was able to study how they played these imaginative games with reality by looking at the videotapes made at each session – with the knowledge and agreement of the participants. They would be trying to imagine they were rain or a bulldozer; they would see themselves as dogs barking or cats purring; they would claim to have magic powers and wonder whether the Indian rope trick could be used for one of their assignments.

The fear of change

But are they worth while, all these efforts to pretend, combine, perceive, all this courage to think irrationally, from the subconscious, and, as it were, in the rough? All this just to find a few more short-lived novelties when the world is already overfilled with products of one kind and another?

The people in charge of Synectics Inc. assured me that they were longing to prove themselves by solving major social problems; but they were hardly ever given anything but industrial and business questions. Unfortunately, only industry and the military were really conscious of the need for new ideas and willing to try them out.

However, this view is not entirely correct. Donald Schon was one of the earliest members of Synectics. He now has his own firm of consultants and advises American public authorities as well as private business. For years he has been studying the obstacles in the way of new ideas. He thinks that, though the business world is always talking of innovation and change, these words are just official jargon to show that they are in step with the spirit of an age continually clamouring for something new. The unofficial but true attitude of management, on the other hand, could be expressed as follows: 'Innovations are dangerous, destructive and uncertain. They are the enemies of orderly planned activity. They turn everything upside down and do nothing but harm. By all means let us speak of innovations, let us study, praise and embrace them – only do not let us promote them.'

How is such a statement compatible with the fact – which is obvious to all – that we seem to live in an era of constant change? We find it so hard to adapt to it that many of us become psychologically or even physically ill. Is the so-called shock of change only an imaginary disease?

For decades we have watched a perpetual firework display of new inventions which flare up briefly only to disappear. But most of them are not fundamentally different from the novelties they are supposed to supersede: they are only variations on an old theme dressed up in new trappings. The unchanging internal combustion engine of the motor-car is symbolic: though it was long since recognized as harmful to the environment, we go on dressing it up in different tin shapes and using it by the million. There has been an enormous number of new inventions since the end of the First

World War, but they have not significantly changed the national, economic and private power structure. Our machinery has been in a state of permanent revolution, but it is still serving the same unchanging interests. The machines have grown stronger, bigger and faster, and can fly higher: but man's lot has scarcely changed. Indeed, if we weigh up our true gains against our many losses, we see that our lives have changed, if anything, for the worse.

American as well as European sociologists have shown that industry's hostility to innovations – as discovered by Schon – always becomes particularly obstinate when the new invention threatens existing structures, either in industry or in the respective rating of various weapons. That is why we find such strong resistance to computers, which threaten the hierarchy within a firm, and against industrial decentralization, which would mean dividing up responsibility. It is even more true when it comes to the long overdue business of trying to find new social patterns. The old order would be shaken. So it is better to leave such innovations well alone.

This conservatism masquerading as progressivism is most obvious in the drive to increase the production of consumer goods while public problems remain untackled. It exacerbates the crises in traffic, in the environment, and in the cities. Profound social changes are needed if we are to have housing that is humane as well as being quick and cheap to build; comfortable and frequent public transport instead of rapidly obsolete private motor-cars; opportunities for real participation; decent working conditions and stable prices. Products should be adapted to their social function, and that should be the paramount consideration. But at the moment they are judged first and foremost by the profit motive. Radical innovations are feared – not only by the bosses who stand to profit most, but also by the workers, who have long since come to accept the social systems under which they suffer. Armaments workers protest strongly against disarmament, steel workers against the abolition of anti-environmental furnaces, car workers against attempts to run down the motor industry in favour of other means of transport.

But since the beginning of the seventies an increasing number of worldwide long-range crisis studies has begun to have an effect. The individual's fears about his own position are giving way to greater fears about human living conditions as a whole, fears which could be assuaged only by long-range planning for social change. Even if there are teething problems, we must find a way to overcome the deep-seated fear of change and clear the road for genuine innovations. Only then can imagination come to the rescue. There are two possibilities: firstly, social experiments might diminish fears about social change by allowing provisory glimpses into possible futures; and secondly – perhaps even more important – man must be educated not to feel his inner security threatened by new social patterns.

Towards an experimental society

At York University, Toronto, Professor G. Hunnius is working with blue- and white-collar workers on the development of 'ideal workplaces'; afterwards the participants try out their inventions in their normal places of work. At Manchester University there is a team shuttling toy cars, buses, ships, trains and planes to and fro in order to try out co-ordinated traffic systems. These experiments were preceded by questionnaires and discussions in which the participants voiced their complaints and wishes. The aim is not to improve individual types of vehicle, but to achieve quick, trouble-free transport by co-ordinating the different possibilities. In a former warehouse in the port of San Francisco, forty young people are taking part in 'Project One', an experiment in leaderless living, i.e., living without orders or rules. They are being supported by the Stern Foundation, one of the most progressive in the US.

These are just a few of the constantly increasing number of social experiments taking place in the world today. None of them is an experiment in the strictly scientific sense, because they cannot and do not wish to use the kind of absolute control that is fundamental to laboratory research in the natural sciences.

That is one of the reasons why, so far, official science has scarcely dared to experiment in this way. Serious social scientists have been unwilling to risk their reputations and positions in this kind of work with its overwhelming mass of data, its uncertain subject-matter (i.e., human beings) and the impossibility of excluding the outside world and its potentially disturbing influence from the experiments.

But the pressure of necessity has been stronger than the desire for perfection. About the middle of the sixties, experimental living projects began to spring up all over the world. Since that time there have been numerous spontaneous attempts to try out completely different social patterns and institutions. People are seeking new forms of communal living, new family structures, new styles of work and jobs, new ways of distributing income. Unfortunately, these projects and manifestations are often known only to the participants and their closest circle; but they all show an active and impatient reaction to society as it is. People no longer want to go on as before 'because that is the only way', and even political struggles for some kind of revolution do not satisfy the desire for more humane living conditions *now*, and not only in the future.

So in the midst of the old order we find quantities of new models. In business, in administration, in local government, in newspaper publishing, in doctors' and lawyers' practices, in kindergartens and schools, Utopias have ceased to be mere dreams: they have become goals to be tested here and now in live experiments.

Many of these attempts fail; they are doomed to fail. Perhaps the participants are not sufficiently prepared, or the experiment has not been sufficiently thought out, or there is not enough money, or the pressure of the ordinary world – indifferent or even hostile – becomes too much. People who prefer theory to practice interpret these failures as incontestable proof that undertakings of this kind are useless. In my opinion, however, even failed social experiments are useful – sometimes especially useful. The history of science teaches us that bold experiments that fail can be far more informative than cautious and more or less successful ones, because more can be learnt from mistakes.

The question is whether human beings should be exposed to almost certain failure. It is true that some participants in failed undertakings may become severely and possibly irrevocably frustrated. But others – and experience has shown that they are in the majority – only mature through such experiences. They will never again accept life as it comes, but will question, examine, suggest and experiment more than ever. They are the forerunners of a more lively, more flexible, more creative society. The imagination did not seize power in 1968, but it has gained tens of thousands of partisans who are already changing the mental climate of the age and will continue to do so.

One might suggest that social experiments should be better prepared and documented than they have been so far, and that if they run into difficulties they should be continued under different circumstances or repeated by other groups. A yearbook or journal to report on failed as well as successful experiments is urgently needed. Communication in the field has just about reached the stage that existed in science at the beginning of the seventeenth century: social experiments are reported by hearsay, travellers' tales, letters, and more or less accidentally through the mass media. None of this is satisfactory.

During a round-table discussion in Paris, Serge Antoine, the publisher of the magazine *2000* and a leading environmental planner in

France, observed that in most industries and in business a product is tested before being offered to the public. But in our so-called advanced societies, there are virtually no experiments (for fear of failure) and very few innovations in the field of collective arrangements, such as schools, hospitals or municipal infrastructures.

Bernard Delaplane, the director of an important chemical firm, added that when a new compound weighing a few grams is discovered in the laboratories of the chemical industry, the next step is production of several litres, after further experiments some hundreds of litres, and finally, if all has gone well, one decides, on the basis of those experiments, to build a factory which produces several million tons a year. But if we look about us at home-construction, it appears that hundreds of houses are sometimes built without any previous trial. Experiments should be made in this area too.

The same observation might be made of most social innovations. When at last (and usually very belatedly) the fear of innovation has been overcome, then the new-born project is sent into the world with no opportunity to feel its way, to learn, to make mistakes, fail and finally develop new possibilities. No: the new school, the new pensions system, the new transport, the new housing estate, the reorganized administration, the restructured firm have to be completely ready and to prove themselves in the hard surroundings of daily life. They have no opportunity to mature by gradual stages.

Claudius Petit, a member of the French '1985 Commission', suggested that all public buildings should be costed 15 per cent too high and that the surplus should be used for experimental building. This idea proved stimulating. Especially in France, discussions about public planning tend to lead to the view that special budgets must be set up for social experimentation.

The Czech pavilion at Expo '67 in Montreal was an example of a social experiment with enormous political influence. In their own country Czech designers, architects and artists had for years been under the thumb of a Stalinist building bureaucracy: now they were free to put their imaginative concepts into concrete form and to express themselves as they wished. The result was so splendid that people in Prague, where the events of spring 1968 were getting under way, began to ask: why do we not feel this spirit in our society? Does our life really have to go on being so austere, grey and monotonous? One of the leaders of the Czech reform movement said to me that the Expo pavilion was as stirring and inspiring for the 'Prague Spring' as the performance of *The Marriage of Figaro* had been for the French Revolution.

It is not only conceivable, but even probable, that experimental prototypes in every field of social life will soon become a regular institution: there will be experimental cities as suggested by the American Athelstan Spilhaus; experimental offices with every kind of new, non-hierarchical organization; firms with a three-day week whose employees will be able to have a second career as a hobby; and perhaps – if the powers-that-be ever learn that even state institutions must be perpetually renewed – experimental authorities, and parliaments, even temporary exchanges of role by rulers and ruled.

Some of these pioneer undertakings, like the new port of San Pedro on the Ivory Coast designed by the French sociologist Georges Balandrier, will continue under observation for many years to see how the new experiment develops. A vast project of this kind was conducted in five American cities (Trenton, Paterson, Passaic, Jersey City and Scranton) with the co-operation of the inhabitants in order to establish how much could be paid to poor families in welfare without taking away their desire to work. No fewer than thirty thousand families were interviewed in order to pick out a representative sample of 2,300 who were questioned and observed in greater detail.

This example, however, shows how social experiments planned and carried out on strictly scientific lines can turn into instruments of control and domination. The experimental society can become a totally planned society unless the following guarantees operate from the start: participation must be voluntary; the participants must be allowed private reservations and full co-determination; and they must be fully informed of the object of the experiment. One could object that it would be better not to become involved in such dangerous developments at all: but the counter-argument is that new social measures adopted *without* prior experiments – or even the total neglect of social innovation – can mean an even greater violation of social progress.

There is a more serious objection: social experiments with state assistance will never have any but the most superficial aims. Given the mentality of our present rulers – in whatever country – that is certainly true. But we must not preclude the possibility of a change of heart at the turn of the millennium, when experiments will have bred an experimental mentality that may extend even to the ruling élites. Here an important contribution could be made by the 'new education' which will be described in a later chapter.

Simulations: games of war and peace

Social experiments are not only uncomfortable and conceivably dangerous for whatever establishment happens to be in power; they also eat up much time and money. The above-mentioned experiment on welfare benefits lasted several years and consumed $3 million. And even that covered only the salaries of the planners, interviewers and evaluators. Yet it was an experiment with only one object, carried out in a comparatively small area. If we want to tackle national, continental or even worldwide crisis problems, then we must invent quicker and cheaper methods.

Efforts to this end have been made for some years. They began in the middle of the fifties and have developed in amazing variety, but they are still lumped together under the heading 'simulations'. They are planning games, a civil version of the war games invented by the Prussians in the nineteenth century in order to try out future strategies in the hope of predicting the outcome of campaigns and battles.

With the invention of computers, conflicts could be translated into mathematical formulae. It then became possible not only to take into consideration far more data about future wars, but also to fight electronic battles much more quickly once the slow and difficult process of model-making had been accomplished – i.e., the representation of all the conflicting human and technical forces by mathematical symbols.

N. C. Dalkey of the American RAND Corporation tells of the model of a war game for two opponents which could be played on an IBM 7044 in a fiftieth of a second from beginning to end. He elaborates:

> At that speed it is feasible to survey literally hundreds of thousands of simulated nuclear wars. If, on the basis of such a rapid survey, it looks like certain strategic attack plans on our part are interesting, then we have a more detailed model which requires about a tenth of a minute to run, and it is feasible to investigate several hundreds of the more interesting cases at that level. Finally, we have a very detailed model that we almost never run, which requires four hours for a single case, and which can be used for a final proving out.

One of the most ambitious American simulation models is called TEMPER (Technological–Economic–Military–Political Evaluation Routine). In the sixties it was used to represent the world powers and their relations to one another. Runs were made to try out every possible alteration in these relationships: conflicts, pacts, declines, stagnation, détente, exacerbation, compromise. Just as in half an hour a planetarium can show the astronomical events of hundreds and thousands of years, so it was proposed to show, on the basis of various assumptions, the possible future historical developments among the thirty-nine most important nations of the world.

At once certain difficulties and imperfections became apparent, the same that were pointed out repeatedly in the early seventies in criticisms of the work of Jay Forrester and Dennis Meadows on the limits of growth. The representation of the world and of human behaviour in computer models is much too schematic, their critics said, and the choice of data much too dependent on the prejudices and values of the model-makers. It is impossible to regard the results of their calculations as objective pictures of reality.

But in the coming years and decades such objections may cease to be wholly true. Even since 1945 the density of data in every field has increased enormously, and it continues to do so. At the same time the capacity of computers to deal with unknown facts is likewise increasing. W. C. Churchman of the University of California foresees future planning games with a hundred million equations. This would make it possible to have several contradictory or complementary, as well as much more accurate, representations of complex realities. It is as though a television picture were to be transmitted with a thousand instead of a hundred dots; what is more, several pictures could be projected simultaneously. A more complicated question is whether it will be possible to express in symbols indefinable, specifically human factors such as joy, happiness, grief and beauty. The mathematician Richard Bellman of Santa Monica and his collaborator Lofti Zadeh think they are on the way to solving this crucial problem. Their work on methods of 'dynamic programming' and their use of 'fuzzy sets' point towards the possibility of programming computers with changing or not exactly definable data.

It will still be impossible to eliminate subjectivity from the design and operation of such mathematical models of the world – and this may be fortunate. Every computer model will mirror the predilections and prejudices of its maker. We shall therefore have to take these representations of reality as modern forms of personal expression, like paintings or books, not as objective representations. The factor of unavoidable one-sidedness is increased by the choice of problems for the computer, and by the decisions made in the course of the game.

Such partisanship is already obvious from the fact that almost all the existing simulations have dealt with war or with the strategy of big business. The position is the same as in the case of technological innovation and creativity research: industry and the military have played a major part in establishing the new methods, and they continue to control the field.

That is why, until now, few peace games or future games have been played on the computers. However, a beginning has been made by Paul Smoker of the University of Lancaster, who has tried to simulate new methods of conflict research, peace-making and peace maintenance. But far greater means and far more qualified computer operators are needed in the field. It is not inconceivable that one day international organizations like the Council of Europe, OECD or the UN will set up peace games on a grand scale. It would then be possible to play off a simulated disarmament without disturbing the economics of the armaments industry. Analysis and simulation combined could work out the most varied alternatives for redeploying credit, plant and labour according to peace-oriented values and priorities. The publication of such alternatives for the future would presumably help to allay those fears about its uncertainty which still block the way for so much essential innovation and change. According to Ernst Bloch, fear of change could then be transformed into concrete hope: the chances of realizing imaginary concepts of a more humane future could be tested to a certain degree by analysis and simulation.

A proposal of this kind was made by Clark C. Abt, the founder and head of one of the leading private consultancies in the US. He threw open to discussion the question of how to stem the American consumption of energy and raw materials. He proposed that the most important industrial consumer goods should be made more durable. Then, instead of eight million cars per annum, the US would need to produce only one million. This would cut the consumption of energy in the subsidiary branches alone of the automobile industry by 80 to 90 per cent.

Naturally such a radical conversion could only be risked if it were preceded by computer studies of the effects on the economy, and in particular on the labour market. Hundreds of thousands would temporarily lose their jobs. Other opportunities would have to be found for them, and for this the understaffed service industries would be particularly suitable. Retraining schemes would have to be prepared well in advance.

It is true that the cost of the improved product would rise. A refrigerator would cost $2,000 – $5,000, a car $15,000 – $30,000. Few people could afford them, so their purchase would presumably have to be financed by mortgages, as house purchase is today. Abt proposes that, in addition, the state should encourage the trend towards durable goods by lowering or abolishing the tax on them. Firms switching production from quickly obsolescent to durable goods should be generously supported.

Abt is convinced that if the advantages of energy- and fuel-saving significantly surpass the costs of the transitory industrial adaptations, it ought also to be possible to overcome the prejudices and habits of consumers. The first step in such a cutback of energy consumption should in any case be a basic analysis of all the costs it would entail.

The firm of Abt Associates in Cambridge, Massachusetts, specializes in studies of this kind. It is run as a private business and currently employs about four hundred young specialists from various disciplines to study social problems. Clark C. Abt was born in Cologne in 1929. In 1936 he fled from the Nazis with his family. He spent a year at school in St Gall, Switzerland, and then emigrated to the United States. Before his fortieth birthday he had found it child's play to set up a million-dollar business. His numerous products are not concrete objects but 'only' studies and opinions, and the word 'play' is to be taken quite literally. At the start of their enterprise, Abt and his colleagues invented a series of 'serious games' intended to make difficult decisions easier for business and administration, and to be used in schools as new dynamic learning methods.

The young immigrant who worked his way through college as a draughtsman, a sandwich-bar assistant and a book salesman wavered continually between his interest in science and technology on the one hand and in literature, painting and architecture on the other. He took his first degree in engineering, but then became tutor in literature at Johns Hopkins University, Baltimore; he wrote his doctoral thesis on poetry and criticism, and lectured on Thomas Mann. After leaving college he worked as a systems engineer for the Air Force by day and by night he wrote short stories, plays and a novel which he describes as bad.

This combination of talents is particularly common among futurologists. They combine the analytical faculties of the scientist with the visionary and formal gifts of the artist.

Systems analysis was invented at the end of the fifties, and it combines exact data and imagination in a unique manner. Large units (systems) like villages, towns, states, businesses or armed forces are analysed in every detail and their interactions are studied from every point of view; then the dismantled whole is reassembled, changing some of the components and using different plans. In order to diagnose systems and how they work, the analyst subjects them to various stresses, as the engineer does his building or the doctor his patients. To do this he finds or invents challenges, problems, questions and dramatic situations demanding solutions.

Abt first worked in this field with a leading electronics firm commissioned by the Air Force to make computer simulations of air battles, remote-control bombardments, space flights, international conflict situations and armament control measures. He was one of the chief architects of the previously mentioned TEMPER simulations – war games for testing anti-ballistic missile systems. The end result of these games always seemed to emphasize the senselessness and uselessness of warlike action, and Abt soon became absorbed in questions of peace and disarmament. He learned to recognize that, in spite of all the technical progress and all the expense, armaments could not guarantee the country's safety; and he began to ask himself how his systematically played 'serious games' could help humanity instead.

That was the origin of the think factory which treats government departments, town councils, school and public health authorities, traffic planners and universities as a doctor treats his patients. There is such a need for this combination of social diagnosis and therapy that the annual income of Abt Associates has increased almost thirty-five-fold in seven years, from $200,000 to almost $7 million. Nearly 80 per cent of this comes from public funds.

Among other things, Abt Associates suggests educational reforms and checks their success; it helps to improve the diet in old people's homes; it deals with the problem of drunken driving, trains Peace Corps personnel, makes proposals for more humane types of housing, develops new anti-drug measures and new unbureaucratic ways of finding work for the underprivileged, whether coloured or poorly educated; it makes plans for a preventative health service, designs better traffic systems and more effective controls for the environment; and it devotes itself especially to the position of the disabled within the 'normal' world.

Abt's young team employs the most modern methods of process research, cost–benefit analysis, low–profile social observation, and systems control; but the 'games' remain central to many of their project studies. They also set the mood of the place. I have never seen more 'playful' or more cheerful offices. Each is gaily decorated and reveals the individual taste of its occupants, ranging from the satirical to the wildly fantastic. They might be the rooms of grown-up children who find boundless pleasure in their activities.

Most of the games are tried out around the conference table before they are fed to the computers. Each player takes the role of one of the parties in whatever the conflict situation may be, and tries to think himself into it. If the problem happens to be a town improvement programme, every interested party must be credibly represented: the mayor, the various political representatives, prospective house buyers, those displaced by the new developments, the estate agents, protesting students and house owners. Journalists interfere, rumours are spread, there are militant minority groups, enthusiastic architects, sceptical planners and – at any rate in the United States – the Mafia.

All this strikes the outsider as a fancy-dress office party. But it appears that games, especially if they are replayed on the computer, produce insights which could not be arrived at in any other way. For the improvised representations of imaginary crises, plans, projects and disasters often reveal surprising but convincing aspects of human and political behaviour. The replays show up possible future developments and complications which may not have received enough attention at the time of the game, and so lead to vital new considerations, concepts and alterations.

In developing these games for civil purposes, Abt was able to lean on an old and originally military tradition. In 1929 Erich von Manstein, then a young general staff officer in the Reichswehr, had begun to play a political planning game with the German Foreign Office. The subject was the critical relationship between Poland and the Weimar Republic. Later on, the General protested that the purpose of these mental manœuvres had been 'to stop us from stumbling into a war that nobody wanted as we did in 1914'.

Among other things, this early planning game showed how difficult – well-nigh impossible – it is to escape from the 'jail of time'. In 1929 Manstein's assumption that the chauvinist Polish army under Pilsudski would be the aggressor still held true; ten years later the situation was reversed, and the aggression came not from Poland but from Germany. Looking back, Manstein confesses: 'We could not foresee that Hitler would break up the close relationship existing between the army and the Foreign Office, nor that foreign policy would one day be in the hands of a man who deliberately wanted to start a war.'

The failure of the political imagination led later 'gamesters' to adopt the motto 'What would happen if . . .', and to try out and play through very bold hypotheses which, at the time of playing, seemed quite improbable. One example was a successful Black revolt which took over the government. It has probably not been without significance for American foreign policy that in the sixties the young Harvard professor Henry Kissinger was very active in inventing and playing political future games. Sudden changes of course, like his later policy towards China, were unthinkable at the time of the cold war; but in the simulated world of the planning games played by Kissinger they were quite normal even at the height of militant anti-communism. Games can be particularly revealing and stirring for the imagination if you take history as your laboratory and historical events for your research material. Certain events are played as they occurred up to a certain point, but then they are changed or even turned upside down: Napoleon wins the battle of Waterloo, Islam conquers the whole of Europe, printing is invented a hundred years earlier or later, Lenin lives an extra thirty years, Hitler is the first to have an atom bomb, left-wing socialists and communists come to power in Germany after the First World War.

Games such as these may seem senseless, even crazy, at first sight, and so they have often been called. But there is no doubt that the participants gradually develop a sixth sense for alternative possibilities and strategies that have been overlooked. They can grasp a larger number of political alternatives, develop a sharp eye for apparently insignificant factors which contain the seeds of future developments, and they acquire a high degree of mental elasticity with the resulting ability to solve problems and make constructive suggestions.

Simulation in the Service of Society – S³ for short – is the name of an inexpensively produced offset newsletter which since 1971 has gone out into many countries from the Californian port of San Diego. Subscribers can see how, from month to month, there are more and more attempts to test systems of every kind by computer – from a simple maize field whose growth is simulated to world problems. There is hardly any type of social organization whose analysis, mathematical representation, and testing under different future circumstances has not been reported by the editors John McLeod and Roland and Joan Werner: new types of city, new traffic systems, decentralized industry and housing, models for the environment, political alliances and international compromise solutions.

Games and their data apparatus are now being used even in areas where the problems have hitherto been thought 'unplayable'. This category includes behaviour patterns depending on psychological factors, such as voting habits, the loyalty of troops, the acceptance or refusal of innovations, expectations of happiness, disappointments. It is now possible to simulate the effects of television broadcasts, the treatment of patients, types of non-authoritarian and non-hierarchical organization, the behaviour of crowds in disaster situations, and even private experiences like dreams. Psychoanalytical models of a personality and its stream of consciousness are fed into the computer: the data may be so numerous that they would be beyond the powers of combination and memory of a single analyst, but the computer is able to devise several possible development aims and modes of treatment for the patients.

Until now man's efforts to represent reality and thus make himself understood have been limited to the use of ready-made concepts. But these new methods enable us to observe extraordinarily complex processes which may be determined by large numbers of interacting data, and to chart their movement and future development. This is a decisive breakthrough in our efforts to comprehend the world dynamically, and the consequences have not yet been fully evaluated. But what we always forget is that, although the computer results look so objective, very dubious values may have got into the programming. The myth of the computer's infallibility has been impressively demonstrated by Joseph Weizenbaum of the Massachusetts Institute of Technology (MIT). He gave his students two solutions to a difficult problem and asked them which was correct. One had been arrived at by a group using nothing but their heads, the other by a computer. Naturally the vast majority of the students chose the computer printout. Later they regretted it, for the professor showed that the second result was based on a deliberate error in programming. The correct result was the one obtained by the old-fashioned method.

Robert Boguslaw, who studied large complex systems and their behaviour at the RAND Corporation and its subsidiary Systems Development Corporation, was the first to warn against belief in the objectivity and neutrality of computer models representing possibilities for the future. He showed that the 'new Utopians' and the supposedly objective systems designers and programmers cannot help letting their work be influenced by the expectations of those who commission them, as well as by their own.

How computerized hopes can be contradicted by reality becomes particularly clear when we look at the war in Vietnam. According to the well-known military correspondent of the London *Observer*, Andrew Wilson, the Vietnam war had been more thoroughly played through, more exhaustively analysed and planned than any other war in history. The programmers had assumed a number of factors concerning the Vietnamese and fed them in accordingly; but the real Vietnamese did not choose to play the role assigned to them. They turned out to be 'somewhat unpredictable'.

Their errors of judgment during the Vietnamese war cost the adherents of social simulation a considerable amount of prestige. But then the Club of Rome commissioned Professor Jay W. Forrester to design his world models and experiment with them. The resulting studies about the 'limits of growth' put the gamesters again on the map.

The Club of Rome tried to examine whether unbridled growth could lead to a collapse of the world system. The global models made for the purpose were much criticized, but they also inspired much constructive thinking. Meanwhile, people have tried to correct the errors and inexactitudes of the early attempts to make prognoses for the future of the globe, and in so doing they have gone beyond Professor Forrester's findings into new territory:

◇ National and regional models for Norway, Canada, Japan, West Germany and the European Community have been worked out with full attention to social and other factors which were neglected by earlier models.

◇ Months before the appearance of the controversial study *The Limits to Growth* (New York, 1972), M. D. Mesarović of Case Western Reserve University in Cleveland and E. Pestel of the Technical University at Hanover had begun a new project entitled 'Strategy for Survival'. It takes into account the main regions of the globe with their different interests, ideologies, aims, decision-making patterns and internal development, dealing with their probable influence upon each other and with their mutual relationships.

◇ That the different world regions are in different stages of development and that they must therefore have different expectations of growth is recognized by the Mesarović–Pestel study; it is also the basis for the 'First Alternative World Model' which has been designed for Latin America under the supervision of A. Herrera in Argentina. This model envisages a reasonable and attainable minimum standard of living as the birthright of every human being, and is chiefly concerned with ways of diminishing the enormous economic gap between the industrial nations of the northern hemisphere and the Third World.

◇ The Dutchmen J. Tinbergen and H. Linnemann have been working on a special project dealing with the expected doubling of the world population within the next thirty or forty years. They are trying to make concrete suggestions for providing enough food, essential consumer goods and services for another four thousand million people in so short a time without damaging the environment.

◇   Two well-known American bodies, the Smithsonian Institution and the Woodrow Wilson Foundation, are studying 'Aspects of Tolerable Growth' at the suggestion of the Club of Rome.

◇   Nobel prizewinner Dennis Gabor, together with several other members of the Club of Rome, is to look into the establishment of a high-level scientific group to decide on research priorities and particularly urgent problems.

◇   Ervin Laszlo, the well-known philosopher, is directing and co-ordinating a study of 'Goals for a Global Society'.

The dangers of an expertocracy

By origin, age and schooling Dennis Gabor belongs to the same group as Leo Szilard. He comes of an upper-middle-class Jewish family in Budapest. He studied physics and engineering, obtained his first degrees in Berlin, and then emigrated to Great Britain. After a brilliant career in research and some particularly fruitful discoveries, he too found that his personal experiences led him to become more and more interested in social questions. Towards the end of the fifties, the magazine *Encounter* (London) published his article 'Inventing the Future'; together with the writings of Ossip K. Flechtheim, Bertrand de Jouvenel and Olaf Helmer it belongs to the pioneer works out of which futures research developed.

The chief argument of this inspiring article was that after the Second as after the First World War the intellectuals failed because they were unable to develop visions of a better future. Gabor has described what he considers a desirable future in several books. They are optimistic in essence, and – again like Szilard – he is convinced that an intellectual élite must direct the development of mankind which is taking a dangerous direction. But he thinks that there is a real danger that, when the tolerant liberals who created this ideology in the sixties and seventies have left the stage, they may be followed by expertocracies, technocracies or even dictatorships.

What control mechanisms are there to prevent this? Where are the alternative institutions that could counter this intellectual autocracy with other concepts? They exist, but they are few; and their representatives – intellectuals to a man – also tend to regard their views as absolute. They belong to left-wing or liberal left-wing groups, and they tend to have very little contact with the population as a whole.

One of their centres is the Institute for Policy Studies in Washington founded by Richard Barnett, Jonah Raskin and Arthur Waskow. It produces new ideas for non-authoritarian schools, for democratic land ownership, for more public participation in the control of radio and television, for the conversion of armaments industries to other products, and for models of worker control. Arthur Waskow believes in putting oases of future living into the capitalist present: there will be conflict, but from it he hopes to draw lessons for the tactics and strategy of change. He catalyses and co-ordinates existing Utopian communes and other social experiments.

At the end of the sixties this institute became the intellectual centre of the New Left in the US. It receives funds from Philip Stern, the heir to a considerable fortune who decided that old-style philanthropy would never change existing conditions and became the financial support of numerous critical drives against the prevailing system in America. He was the first to finance Ralph Nader's successful campaigns for the consumer. It was Stern who paid the travel costs of Seymour Hersch, the reporter who revealed the horrors of My Lai, so that he could get the necessary evidence. He also supports a number of communes in their attempts to find and test new life-styles.

Stern has long ceased to be the only millionaire who has used his income as well as his fortune to help create an order in which there will be no millionaires. In San Francisco, six young women and seven young men have joined together to support 'constructive changes in our society'. They regard their contributions not as philanthropy, but as self-imposed taxation. George Pillsbury, heir to the Pillsbury Flour Company, wants to go a step further: he has announced that he intends to use all his money for five years to support 'grass-roots community systems committed to making fundamental changes in the system'. Together with nine other scions of wealthy families, he has founded the Haymarket Foundation, named after the famous Haymarket riot in Chicago. 'Change, not charity' is the motto of this group. From time to time they publish a newsletter, in one issue of which this statement appeared: 'Social change is not a window dressing. . . . It can only come from actual change in the structure of power.'

These rich radicals are deliberately – and without any illusions – cutting the ground from under their own feet. They want to plant something completely new. They aim beyond all the experiments supported by industry, the state, and other foundations, which seek no fundamental change, but at best to improve the existing fabric.

Indeed, our world is imperilled by the rush of 'progress' in the name of principles which have become dangerous. We need to rethink radically, to take enormous risks, to question our own position. We need self-criticism, not laments.

Environments for social change

How and where can ideas for decisive new directions develop? In the Eastern bloc? I have had many opportunities to speak of these problems with communist futurologists both inside and outside their own countries. Some of them admit, in the strictest confidence, that they have little opportunity to develop new social concepts and institutions. True, the Romanians have their mathematics centre under M. Botez in Bucharest, where they play with all kinds of future combinations and even develop 'eccentric' design. But little of this is told to the ordinary Romanians. The Polish researchers have been allowed to develop a few models on leisure and culture. But other institutions in their society, which are far more in need of change, remain taboo. In the Soviet Union 'social prognostication' is highly developed; but it is only allowed to analyse, not to make suggestions. 'Our rulers won't have it,' a prominent futurologist said to me in Moscow, pointing heavenwards.

Only in Czechoslovakia, during the short period of the 'Prague Spring', were any really new models developed for a more humane type of socialism in many sectors. At that time R. Richta's group was supported by the Academy of Sciences, and as the régime became less rigid they developed several ideas for a more humane organization of work, for real participation in government by the citizens, for widening the cultural base, and for a cybernetic society which would constantly observe the results of the measures it took and alter and correct them in accordance with this feedback.

In April 1968 I attended the last congress of this intellectually active, imaginative and undogmatic group at Marianske Lazne, formerly Marienbad. I remember how Ota Klein – I can reveal his name because shortly after the invasion by the Warsaw Pact troops he lost his life in a car accident abroad – told me, as we stood outside the dilapidated old Kursaal where the congress met, about the difficulties facing the organizers: 'An ageing, petrifying avant-garde is almost worse than the forces of reaction. They at least have a bad conscience and try to amend their worst faults. But those people', and he pointed to some of the participants, 'think they are the pioneers of mankind, even though they have not added a single idea to the Marxism of their grandfathers' day.'

During the last session, one of the Western visitors made a special appeal to the Soviet participants and begged them in their own interest not to oppose the spirit of social experimentation that was beginning to flourish in Czechoslovakia. Once the friends of Czechoslovakia had feared that the country would be invaded by troops from the West; but now, in April 1968, the danger appeared to be coming from the East, he said.

During the final celebrations one of the Soviet experts demonstratively embraced the chairman in order to show that such defamatory misgivings were quite ridiculous. Then he kissed his Czech colleagues on both cheeks. Today most of them are dead, or have emigrated abroad, or vegetate, stoking boilers, washing corpses or doing odd jobs.

For a long time it was the Yugoslav communists who had the best opportunities for developing the social imagination. The 'Praxis' group published work in their widely distributed journal; there were annual philosophical congresses on the Dalmatian island of Korčula, and biennial international conferences on 'Science and Society' at Hercegnovi. All these provided a forum where the future of the world could be discussed openly and imaginatively. But in Yugoslavia too, social imagination becomes more shackled every year.

In the introduction to this book I spoke of the crisis situation at the turn of the millennium. To deal with the difficulties of mankind we need new concepts and ideas not only in science and technology, but especially in the social sciences. But so far experiments in this field have scarcely ever been financed or tolerated unless they seemed acceptable within the existing power structure.

Is there a way out? Can we, under existing circumstances, envisage creating institutionally guaranteed 'free zones' in which more radical ideas can be tried out – preferably with the participation of those concerned?

There are beginnings, and they could be developed if they had support. The obvious move would be to entrust the universities with the independent development of social projects, for in many countries they still enjoy relative freedom from economic and ideological pressures – though just recently these pressures have again begun to increase. Still, the universities provide a reasonably free place where young people with a relatively high capacity for original thought can meet older specialists with considerable factual knowledge and experience. Ideally there should also be time to consider, plan and experiment, so that the universities do not turn into broiler factories for specialists. It is no mere chance that I have qualified all my adjectives of approval in this paragraph: in reality the universities are not as free and open as they should be, nor the students as creative, nor the teachers as well informed and critical, nor time as plentiful as their founders must have hoped.

All the same, the conditions for creative work are still better there than in most other institutions, if only one knows how to use them. And there are universities that do know how to use them:

◇   At the London School of Economics, Professor Thomas Marcus and his students have developed new forms of participatory education and administration.

◇   At the University of Bielefeld, Hartmut von Hentig and his pupils are working on new types of schools and curricula.

◇   The Universities of Oslo and Bergen are testing models for genuine workers' participation.

◇   The University of California at Los Angeles has started a 'Creative Problem Solving Program', with at least six to eight courses a year, under the leadership of Marvin Adelson. The participants not only work on specific problem-solving methods, but also pursue such unorthodox educational aims as 'Accept uncertainty'; 'Understand the dynamic tension between stability and change'; 'Work with and for real people'; 'Develop strategies for social change'.

These things are going on at a few universities, and they could spread to others. There would then be project universities as well as teaching universities. They would be able systematically to foster the creative faculties of students, which mostly lie fallow now, and direct them specifically into social planning. Learning would not suffer. Jean Piaget, the great developmental psychologist, made the following pronouncement as a result of his studies: 'We learn most when we have to invent.'

Planning should be undertaken by students of all faculties working together. It would be a splendid way of trying out the interdisciplinary studies that are always being demanded. For instance, when William Seiffert (MIT) was planning a mass transport network (the Glideway system) for the American Northwest, he employed not only every kind of technical specialist, but also economic geographers, statisticians, demographers, sociologists and political scientists. Working together but each in his respective field, they tested the new traffic system, its preconditions and its consequences.

Another environment where social planning is beginning to flourish is provided by certain foundations and the institutes they finance. In France the 'Fondation Royaumont' has initiated research into human development; in Germany, the 'Carl-Backhaus-Stiftung' in Hamburg is researching into and promoting more humane and democratic conditions of work; in England the Scott Bader Foundation is trying out new forms of workers' co-ownership. The 'Max-Planck-Institut für die Erforschung der Lebens-bedingungen in der wissenschaftlich-technischen Umwelt' (Max Planck Institute for research into living conditions in the scientific and technological environment) founded by Carl Friedrich von Weizsäcker in Starnberg tries to develop models of alternative economic structures and new technological systems. Alas, even that socially minded think tank takes little notice of the general public – which after all supports it with its taxes.

The same reproach could be made to almost all progressively minded universities and foundations: they are isolated from ordinary citizens, from the people with whom they claim sympathy in speech and in writing. Why? Presumably not for fear of seeing their privileges encroached on, nor from a sense of their own superiority, but because they believe – although they do not often say so – that only people who are 'properly qualified' can usefully take part in these intellectual preparations for changing and improving the world of the future.

The new social barrier of qualification is nowadays often guarded much more strictly than the barriers of wealth, status, race or class. Some of the old areas of reserved privilege have been thrown open to all: many palaces and gardens are accessible, at least occasionally, to everyone. But just let an ordinary citizen try to look in on a congress of specialists without being invited; the stewards will soon throw him out. It could happen even at meetings of the trade unions' intellectual élite. In 1971 Germany's biggest trade union, IG Metall, held a conference in Oberhausen on 'Tasks for the Future'. It cost about DM 1 million, the money naturally being derived from membership fees. But the building where the functionaries and their guests met was as closely guarded against members without invitations as if it had been a high society gala.

This attitude reveals intellectual arrogance, complacency, and most of all the erroneous conviction that it is impossible to have a sensible and fruitful conversation with the man or woman in the street. Sociologists say that the 'cultural barrier' effectively prevents the masses from taking part in debates about their own lot. They are considered too ill informed, too indifferent, too far behind whatever the present state of knowledge may be, uninterested and uninteresting – in other words, those crippled through no fault of their own are blamed because they limp. It is true that the vast mass of people in this so-called age of information are far too poorly informed. But does that mean that they have no experiences, no thoughts, no ideas of their own? That they have to remain silent because they have nothing to say?

The silent shall speak

How wrong – or at least partially wrong – this attitude is was brought home to me when I visited the Italian social reformer Danilo Dolci in Sicily in 1956. One of his friends met me at the boat, because I was to join him in one of Palermo's *bassi* where it is supposed to be dangerous for foreigners to set foot. Dolci was there on a hunger strike in order to draw the Italian public's attention to the need to clean up these dilapidated quarters which were overrun with rats and stank of excrement. I found him in an empty shop open to the street. There he lay, visibly weakened, and talked to the people who crowded around him in silence as though they were adoring a martyr.

'There, now you've seen me. But I'm not a saint – I'm just one of your fellow-citizens who unfortunately got too fat,' he laughed. 'Let's have a talk, and I want to ask you a question.' They nodded their heads and waited. There were young and old, men, girls and children, and in the sharp white light of the neon lamps you could see that they were all marked by the poverty, the endless misery, the sadness that is so common in the Italian 'deep south'.

'Well, tell me what you expect from life?' Silence. A few dirty jokes from the back row, then not another word. Dolci too was silent. He remained silent until the silence became torture. Then at last he said: 'But you can talk to me. I'm not a boss, I'm not a priest, I'm not a learned man. You are my friends. Tell me, what do you expect from life?'

Still no one dared speak, so he had to point like a teacher to someone in the front row. He pointed to a second person, a third. The fourth was a lean man of about thirty who looked much older, and hesitatingly, almost stammering, he began to say something. A fifth man joined in, a sixth continued, and soon they were all talking loudly about what they wanted and had not got, and what could be done to get it. The interpreter whispering in my ear could no longer keep up. 'He says, they ought to bring the olive trees into the town. And food should be cheaper. And that one: he wants to abolish money altogether; he has a plan to find work for everyone. And the madman over there, he wants the *carabinieri* to take off their clothes and patrol around in the nude.'

It went on for two or three hours, a volcanic eruption of hope and hatred, of confused wishes and largely impossible ideas. The mute had learnt to speak. The desiccated, unused brains began to think. They all had imagination, and they were glad to offer their ideas: ideas that were hardly feasible, ideas that ought to have been carried out long ago, and ideas that needed thinking about because there was something in them.

I learned a lot on that unforgettable evening. The silent can talk if they feel that you are listening. They have the courage to overcome their inexperience and shyness if they meet not with contemptuous superiority but with a brotherly fellow-human who does not regard what they have to say as inevitably stupid, ridiculous or primitive, but as the valid expression of a unique personality.

The so-called 'future workshops' which I run for many different groups have taught me how much unacademic people in particular feel the need to express themselves and produce their own ideas. I hold these sessions with apprentices, manual and office workers, agricultural labourers 'and unemployed people. We start with a complaints period: we bring up our anxieties about the whole range of problems to be discussed and table them in large, easily visible writing on long rolls of paper. In the Revolution of 1789 the French used a similar form of criticism which they called *cahiers de doléances* (bills of grievances).

But with us, criticism is only the prelude to constructive proposals. In the next round we use the brainstorming technique to produce ideas, hopes and desires that may perhaps lead to improvements. *Ad hoc* groups are formed to develop the most interesting ideas and visions in greater and more concrete detail. In the next phase of this exercise of the social imagination, these models are submitted for criticism not only to the group members, but to experts as well: their originators have to defend them. In the last phase, the social inventors and the critical realists combine to examine the chances of success as well as the obstacles which their proposals might expect to encounter in the world of reality. Together they plan a strategy for putting their ideas into practice, or else reach a clearer understanding of the forces opposed to their realization: they end up more strongly motivated than before to seek political solutions.

Our efforts to increase the potential resources of social creativity are only in the earliest stages. The first experiments have produced an interesting and unexpected experience. People who, on their own admission, either no longer took an interest in political and social questions, or never had taken such an interest, were led by these sessions to seek information about community matters and to become politically active. Why? Because at last they had taken part from the very beginning, had progressed from criticism to formulating problems and finding solutions, and finally to confronting these solutions with facts that had hitherto been unknown to them. A participant in one of the 'future workshops' once expressed it like this: 'I never used to be interested in children. But when I had one of my own I began to watch children more closely and to think about them. The idea I put up for discussion just now is something like my baby. And now I want to know the best way to bring it up.'

I must admit that the number of really new ideas produced in these socially oriented brainstorms was small. But that is not surprising. Long years of neglect and frustration have severely damaged our contemporaries' active creativity and faith in their own powers of invention. It will be necessary to unearth long-buried faculties in every individual, especially in those who are deprived of any real rights of participation. What we need is an 'Everyman Project', and I shall now describe the beginnings of it.

2 Everyman Awakened

Poverty in wealth

Luigi P. attacked the precision machine with both his fists. For six weeks he had been punching out micro-circuits on it. He howled like a maniac, his workmates reported. No, more like a mourner, said a would-be wit: the way that Neapolitan women still mourn by throwing themselves screaming on the coffin.

But why argue? The painful incident is as good as forgotten. The personnel officer talked to the man. The works psychologist had a few sessions with him in his gleaming white consulting room, and now Luigi is back in front of his instrument: a blue-clad, well-groomed model worker in Olivetti's model factory above Pozzuoli. I saw him working there. He worked his machine with precision, and with the same precision he answered the questions of the public relations man who was taking me round the factory; he was a contented workman and well paid by southern Italian standards. His breakdown is described to visitors in a slightly lowered voice: it has become an interesting peculiarity which lifts him out of the crowd and makes him an example of enlightened modern factory management.

This branch of the great Italian-American electronics firm was built in the sixties and is regarded as one of the most modern factories in the world. From the sea it looks like a palatial private property with its snow-white façades, mirror-like windows and tropical gardens. The architect, a Neapolitan communist councillor, was accused of un-Marxist behaviour by his comrades because he built such a beautiful and progressive factory for a giant capitalist concern. His counter-argument was that it was wrong to wait for the revolution: workers should enjoy the best possible conditions here and now. Once they had experienced working in a specially privileged environment, they would be inspired to fight all the harder for more humane conditions of work for all.

But the building was criticized even on the drawing-board by the factory managers and engineers. It was no proper factory, they said; more like a casino or a big ballroom. It might be all right for parties, but not for serious precision work. But Adriano Olivetti, the head of the firm at that time, was not to be diverted from his purpose by these criticisms. He was an obstinate, sensitive and benevolent man who had emigrated during the fascist years and had returned with the intention of creating something new. His newest plant was to be built according to the latest principles of ergonomics – the new science that tries to reduce the mutual wear and tear between men and machines.

Therefore no workplace was to be more than twenty yards distant from a tree. The windows would open and shut automatically at certain temperatures (a factory that breathes, according to modern builders' brochures). The height of the seats, the distance between individual workplaces, the convenient arrangement of handles, dials and controls on the machines was all minutely measured and calculated. And that was not all: an artist was commissioned to provide constantly changing interior decorations, to give art lessons to the workers, to organize exhibitions, and to make portraits of every member of every team at work in order to make them feel that they mattered as people.

Of course these things cost a lot of money. But the high productivity rate in the new factory – the annual average is 4–6 per cent higher than the average of the parent factory at Ivrea in northern Italy – justified the boss's idealism to the most calculating members of the management. Besides, there is always enough labour in southern Italy; it only needs to be trained. And as the people are lively and intelligent, that is not too difficult.

So everything seems to be going nicely. Wages are above average for this part of the country, and the number of applicants for jobs correspondingly large. Only people with special recommendations – from a priest, a schoolmaster, or a business contact of the firm – can hope to be chosen (one might almost say elected) from among the plethora of candidates. When a man is taken on, the event is celebrated as though he had been elected to an English Club or had gained entry to a higher circle of society. And in fact a Pozzuoli Olivetti worker brings as much honour to his family as if he had become a priest. Only in this case he is a 'holy man' who brings home money to boot, a person who can be used as a reference by his relatives, can even be approached for a loan.

That, in fact, is the trouble, said the personnel manager as he tried to explain the Luigi case to the foreign reporter: the parasites, the spongers, the false friends, the impoverished aunts, the marriageable sisters, the nephew who has got into debt, not to speak of the old father or grandmother, all clinging like limpets to the poor model child of the family. And on top of that there are his own new obligations: the hire purchase repayments, the repairs, petrol, clothes for the wife, the children, for himself: 'Poor devil.'

The works psychologist knows better, but must not say so out loud: the truth is that it was the work that drove his patient to despair. Yes, his work in this showcase of factory reform. 'Bloody kindergarten,' Luigi screamed as he smashed his toy.

What had his life been before? Not many *lire*, but a lot of fun. He repaired television sets, got cars back on the road, went fishing. He had been fiddling with an outboard motor when he recognized his talent for 'healing machines', and soon everyone with mechanical problems came to the 'wire doctor'. He had helped out as a waiter, driven trucks in Sicily during the orange and lemon harvest, done a newspaper round – and in the distant past he had even sold smuggled cigarettes. Life was never boring. He had not lazed about much, but he had chosen his work as far as he could, as well as the speed at which he did it, and had decided for himself that he enjoyed it or at least did not find it too hard and troublesome.

None of this in the model factory. You are trained. You repeat the movements you have been taught. You write down what has to be written down. You get your daily quota and then you get your wages according to the scale laid down by the union. Show that you are worthy of it. Have a sense of duty about fulfilling your norm.

Millions the whole world over go through this adaptation process every day. Very few complain. They may make a few jokes about their 'shitty' but not intolerable daily life. They are told to fall in, and they fall in. But Luigi got out of line. They calmed him down and when his rage was spent they magnanimously forgave him. When he did not want their forgiveness, they tried to coax him. When he would not be coaxed, they pointed out – in the friendliest manner possible – what a flood of debts could engulf him; they frightened him with anxiety about his poor family and himself; perhaps his outburst was the beginning of a serious illness which needed to be carefully watched . . .

He told me all this himself after work – for of course he still works at Olivetti. He has toed the line. He is even politically organized. His trade union branch secretary tells him when his subscription is due, what he is to say in discussions, when he is to turn out for demonstrations. A poor devil – that is what he says about himself in his more lucid moments. But the people outside envy him – they envy him even after what he has been through.

Luigi is an example of an unsuccessful revolt against alienation. Many workers experience similar temptations, but mostly they remain wish-fulfilment dreams. West Germans, Belgians, Swiss do not boil over so fast and are quicker to mobilize the police forces of a false conscience that comes disguised as good sense. But all of them suffer injuries to their personality. Posters warn against the danger of physical accidents: a whole system of safety precautions has developed in the wake of technological progress. But injuries to the inner man go unrecognized. Sentimentality! Self-pity! Stupid talk! Even the victims upbraid themselves in these terms. Silence reigns over the psychological abrasions, festering sores, calluses, mutilations; people repress their memory of how freely their imagination once flowed, they anaesthetize the pain of losing their selves.

Harvey Swados, the late American novelist who had worked in a car factory, wrote from his own experience that, almost without exception, the men on the assembly line felt like 'animals in a trap'. They were 'sick of being pushed around . . . sick because they had to work like a donkey with blinkers, sick of their dependency . . .'

This description was given in the year 1957. Not much has changed since. I have visited several car plants. It may be that the physical conditions at the conveyor belt in the Ford factory in Detroit where Swados worked have improved – the lighting is better, there are places to sit down and rest – but the belt goes on moving, and faster than ever, especially in the most modern semi-automated plants such as the General Motors works in Lordstown, Ohio. What remains unchanged is the fact that it is impossible to develop any scrap of judgment, skill, imagination, or even motivation. The workers go on taking orders, they follow instructions, they obey, they remain the tools of another's will, they are passive.

It is characteristic of today's situation that the number of people taking orders is on the increase, while the number of people with even a limited scope for making their own plans and decisions is declining. There are plenty of studies, statistics and committees to counter the concentration of the economy, but no surveys about the growing concentration of creative activity and autonomy in the hands of a shrinking circle of people. And no efforts are being made to combat this evil.

So a new sub-species of have-nots is developing. They have decent wages and savings accounts, and they live in new houses full of consumer goods – but they are still impoverished. For neither the savings account, nor the pension, nor the free weekend, nor the annual packaged tour can alter the fact that they have been deprived of the chance to create something of their own, to exert their personal influence, to give the world a testimony of their own special personality. They do not live – they are lived.

They find their compensation in consumption, in aggression, and in sex – the last bastion of personal development, but now also imperilled. Many lead a double life. The tiler passionately collects tropical fish and spends a few weeks each year fishing in the tropics; the man with the electrical shop goes on exploration trips; the machine operator captures birdsong with his audio equipment. An office worker on holiday at the 'Club Méditerranée' in Corfu told me: 'For eleven months of the year I work like a machine in order to have one month living like a human being.'

People rich by the standards of self-realization often find it hard to make ends meet: people such as artists, gardeners, students, young scientists, craftsmen, independent film-makers, all of whom lead an eventful existence. It cannot be denied that they too have to worry about supply and demand. It is true that they too are led by the minority with the material power, but on a longer leash. Yet even the little extra bit of initiative and self-determination that they have lifts them above the great mass of those controlled from above.

Marx saw the masses deprived of their personal creativity, and their resulting impoverishment in the midst of a world of external plenty: he regarded it as the central problem of capitalist industrial society. But the struggle for better material conditions for wage-earners pushed aside the striving to emancipate the personality, which was banished to the periphery of Marxist-inspired social criticism. It was assumed that a change in the economic power structure would eliminate the workers' alienation from their work and its products. It was hoped that when they received their rightful share in the ownership of the mechanical tools of production, they would also resume ownership of their own individual tools of production – i.e., of their native individual creativity and the opportunity to use it. But this hope turned out to be deceptive. Over the last twenty years I have visited offices and factories in all the European socialist countries and have nearly always found the same inner indifference, the same resignation or unspoken though evident dissatisfaction, as in similar places in the West.

I use the qualification 'nearly always' because I found two exceptions: one was a hotel at Hercegnovi in Yugoslavia which was largely run by the employees themselves; the other was a workshop in the Soviet atomic centre at Dubna. In the first case, the difference may have been due to a genuine right of participation in decision-making; in the second, it was because all the centrally produced parts that were delivered to the assembly shop in this great international research centre had to be specially altered and adapted for certain experiments. That was why all the workers showed the intense personal interest that one associates with people engaged in their chosen hobbies. One of my researcher friends told me that they developed positively personal feelings for some of the highly delicate and precise instruments which often arrived in poor condition and had to be rehabilitated.

Imagine that only a small privileged group in society was allowed to produce children. The rest of mankind would be condemned to sterility. That gives you an idea of the present situation with regard to man's creative potential. The mental sterilization of millions of people was carried out in the cause of so-called 'higher development'. Everything around them grew ever more efficient, more perfect, more powerful; they alone became more and more dependent and powerless in the process. It is inconceivable that people will continue indefinitely to submit to the increasing usurpation of their opportunities for self-realization. More and more of them are no longer willing to find themselves psychologically and intellectually impoverished for the sake of external prosperity – which in any case is often somewhat doubtful.

This tension may – yes, one can predict that it will – lead to a counter-movement whose members will not be bought off with higher wages, nor with more consumer products, nor even with the kind of reorganization that would give them nominal co-ownership while leaving them dispossessed of their personal creativity and of the power to influence decisions. They will demand participation in what is produced, what is decided and what is done. They will demand the liberation and development of their creative potential instead of its further atrophy.

The revolt of the awakened

In the last few years, more and more cars have been coming off American assembly lines with damage by sabotage. Brand new machines have been found with badly or incorrectly riveted joints, with bent parts, or with tears in the upholstery. Not infrequently as much as a fifth of the day's production has had to be rejected by the checkers and sent for repair before it had even been used.

Investigations have shown that those responsible for this sabotage were mainly young workers. When they were challenged and appeals were made to their 'better education', which should have deterred them from 'such senseless acts', they responded by cracking jokes. Later on, more careful investigation showed that it was in fact the slightly superior degree of education among this new blue-collar generation that had caused their permanent protest against the monotony of their work, which they regarded as an insult to their intelligence.

But there is another, more profound motive. 'Every one of those things has my mark on it,' an intelligent young mechanic told me with a mixture of pride and contempt. This was at the Detroit Industrial Center of General Motors – one of the model factories of this giant organization, designed by the Finnish architect Saarinen with fountains and ornamental lakes reminiscent of Versailles. I had been introduced to this worker because he had distinguished himself in an essay competition on the subject of 'My Place of Work'. His prize-winning essay probably did not mention that he scratched his 'coat of arms' into the freshly sprayed paint in places where it was not immediately noticeable, that he made small dents in the body-work, or that he sometimes reconnected the leads so that the button for sounding the horn would set the windscreen wiper going. Practical jokes of this kind, so investigations have shown, are in fact the expression of a very serious need for self-realization.

Some time ago the dingy carriages of the New York subway suddenly began to sprout huge black inscriptions, blue stars, green jungle plants, yellow snakes and other wild paintings. At first they were thought to be the work of a single unknown individual who used felt pens and spray-guns to write his nickname, 'Taki', and his address, 183rd Street. But soon he was joined by other amateur artists: Spin 70, Hondo 127, Snake 13, Super Cool 119 and Stay High 148 smothered the trains with a fairy world of fantastic figures and shapes. The authorities tried in vain to stop these so-called acts of vandalism. Heavy punishments were imposed for disfiguring the trains. But the revolt against ugliness and anonymity continued. At depots and at empty stations by night, hundreds of graffiti were written, painted, sprayed and engraved, and the dreary grey background of a daily life of dependence and creative impotence was transformed with hundreds of personal marks.

The turn of the millennium will probably be characterized by a continual rebellion of the more talented whose newly awakened potential finds insufficient expression, or none at all, in the world of work. This is the result not only of a growing loss of the right to be heard, but also of a growing realization that mankind is being deprived of its most personal possession. In schools, on college courses, and by means of the mass media, more and more young people are beginning to develop their critical understanding, together with a longing for creative activity. The old aims of the Enlightenment could now come closer to realization on a much wider front, but it is becoming apparent that society is not ready for millions and millions of bright minds and continues to employ most of them at levels below their capacity.

The rebellious car workers on the assembly line, the dissatisfied steel workers, the dial-watchers in hydrogenation works who develop neuroses, all these are symptoms of the gap between people's potential and actual performance capacity which can be observed in many other fields of employment. Workers in the service industries do not have to struggle quite so hard against boredom and routine as those in the highly rationalized production industries, but even people in slightly less monotonous jobs suffer increasingly from lack of self-realization and the apparently senseless nature of their work.

In the twentieth century new mental powers have been developed by education and information, but we have not understood how to absorb and use them any better than we did earlier discoveries of material and physical possibilities. The reaction to this crisis follows a predictable pattern: there is a demand either for running down and curtailing so-called excessive educational provisions, or for more control over training. Instead of opportunities to develop, people are to be offered precisely defined roles within carefully programmed organizations without any say in their design, construction or control.

Must we have an academic proletariat?

The cry of panic: 'Stop the flood of academic qualifications!' and the increasing limitations put on entry into higher educational institutions in the industrial nations will remain unavoidable as long as we have a society incapable of using the growing supply of highly qualified people because, in the words of the Munich industrial sociologist Lutz, 'the employment system does not correspond to the education system'. However, many of his colleagues feel that more suitable occupations for the increasing number of people with higher qualifications can and must be found.

But the problem will have to be tackled from another angle as well. In our modern performance-oriented society, only certain types of ability are valued so highly that their training confers social prestige. In terms of the population as a whole, careers of this kind offer few openings and the struggle for them is correspondingly intense. As a result, the process of selection begins long before adulthood, and wide sections of the population are disadvantaged, either because the system does not cater for their particular abilities, or because it fails to reawaken and cultivate abilities already impaired as a result of prejudicial social circumstances.

One of the most admirable achievements of education in the socialist countries is the attempt to eliminate early deprivation. But with the exception of the Chinese method, even these school systems neglect manual ability and the special emotional sensitivity and high degree of originality which can manifest themselves as disorderly behaviour. Even there, education proceeds according to narrow and largely middle-class concepts of what behaviour and intelligence should be.

One of the most encouraging changes – and we shall be able to see the results better in a few years' time – is the intensified effort of educationalists to reject the criteria of former generations with their discrimination and neglect of human potential. New beginnings in this field, prematurely though perhaps understandably decried as Utopian by those with privileged positions to defend, deserve much more notice and encouragement from society as a whole. What is needed most of all is a little more patience and more financial support. Most educational experiments are interrupted for lack of persistence, lack of money, or lack of sufficient encouragement from outside. Apart from that, the people engaged in them waste far too much time criticizing other reform experiments. They are usually more interested in defining the differences between themselves and other challengers of the old system than in recognizing the elements they share with other innovators. We need more tolerance not only from society, but from the 'change agents' among themselves.

As I see it, the Everyman Project should not be centrally controlled and co-ordinated in every detail, like the Apollo Project; it should be a loose framework for trying out various methods of self-realization, a sanctuary for human development and change not merely tolerated by the world at large, but positively encouraged.

The Apollo Project had to overcome the pull of gravity; the

Everyman Project will have to contend with the enormous weight of anxiety and prejudice among all those who think that it is impossible to liberate human ability on a vast scale – or who fear such a possibility. They will, of course, disguise their fears as rational argument. But not so long ago, so-called realists were decrying as fantastic the idea that man should be able to fly; that he would be able to crack the nucleus of matter; that he would be able to reveal the mechanism of genetics and life itself in sub-microscopic experiments. In my opinion, people who cannot imagine the upward development of vast numbers of human beings are equally short-sighted and unimaginative. They regard the undeniable failures that have occurred in the field of mass progress as proof positive that such a thing is impossible. But these early failures ought, on the contrary, to be regarded as challenges to make fresh and ever more imaginative and intensive efforts.

Counter-cultures today

It is well known that astronautics began long before the first moon landing. Similarly, the beginnings of the Everyman Project go back many years. The Weimar Republic, and Vienna in the decade after the First World War, were especially fertile in experiments in educational reform and mass education. Many of the experiments failed, but they scattered new stimuli like seeds from a pod. Currents of information bore them in all directions, and they took root in the most unexpected places.

As always, the rulers tried to control these new forces and to use them for their own purposes. But they were only partly successful: the demand that each individual should have his own judgment and his own skill was not suppressed but reared its head in ever new experiments. Surveys like the one undertaken by the Organization for Economic Co-operation and Development in the early seventies show that there are hundreds of experiments in schooling all over the world. Some last only from one to three years. Money runs out, there is too much opposition from outside, or too many internal difficulties and disagreements. But each failed new school is superseded by several others. What is even more important is that books, journals, teachers and pupils spread the experimental experience among the public and into conventional education.

A survey among English head teachers showed that 'the father of anti-authoritarian education', A. S. Neill, had no direct influence on the curriculum of the state schools. But so many teachers, parents

and children absorbed his ideas that the whole educational climate
in Britain has changed somewhat even in state education.

There are tens of thousands of educational establishments the world
over whose main efforts go towards adapting their pupils to the
existing state of affairs; but there is also a sizable number of experimental schools, education shops, project groups, seminars, conferences, student meetings, encounter groups, and exercises for making
educational counter-culture a reality, and all these attempts have an
influence out of all proportion to their numbers.

In the midst of the urban reality of rigid building complexes and
regulated traffic streams, adventure playgrounds are springing up;
even in primary schools, children are organizing themselves against
state and parental oppression; children are being educated not for
merciless competition, but for mutual aid and friendship; adolescents
are being encouraged to have faith in their own ideas and dreams
instead of being trained merely to carry out instructions; in a world
of scientific over-specialization, students are being allowed to study
wider relationships and interactions; even the man in the street has
been admitted into the world of knowledge through the gates of
unconventional and progressive education, for instance at Bremen
University or through the Open University of British television.

One cannot help suspecting that these 'open spaces' are only
permitted in order to divert and control the longing for change.
They might be no more than alibis or fashionable disguises for the
old conditions. It is quite possible that crafty calculations of this kind
have led powerful decision-makers into granting concessions which
they present to the outside world as democratic tolerance but to the
initiate as tactical ruses. Be that as it may; one way or another these
changes have become inevitable in the face of historical developments
which can no longer be totally ignored or suppressed.

The Soviet research centre of Dubna contains not only impressive
experimental installations for physics, but also schools with a degree
of pupil participation unthinkable elsewhere in that country. There
a leading official told me that in the Soviet Union schools, universities, laboratories and in fact the whole field of culture are to be
regarded as 'sectors of continuous change'. Was he deceiving himself? At first sight it would seem so, for socialist countries are only
too liable to condemn innovators as revisionists and counter-
revolutionaries. And yet he was not quite wrong. But in the peoples'
democracies the powers of control and prevention are so strong that
any movement is scarcely perceptible from outside.

Today this isolation is one of the worst handicaps of the socialist countries. For, contrary to the opinion of protagonists of the working-class struggle, the contest between the capitalist and socialist systems will be decided not according to production figures or the concentration of power, but by the degree of openness and receptivity to innovation and change shown by every citizen.

The new schools in all their varieties are unrealistic only in relation to the *status quo*, not if you look at them from the standpoint of future development. From that angle, on the contrary, we must concede them a high degree of realism. For – possibly without being aware of it – they have developed a fine instinct for the special demands that will face mankind at the turn of the millennium:

◊ independent judgment, imagination, insight and foresight to deal with a complex situation of individual and collective crises;

◊ mental agility and adaptability to deal with swift changes;

◊ tolerance and solidarity to achieve peaceful co-existence in the age of the population explosion.

The number of pupils being educated in pioneer schools or under progressive teachers, or taking part in other new educational experiments such as communes and co-operatives, is comparatively small, but that is less significant than their general influence. Every innovator carries hundreds of indifferent individuals along with him. The new teaching methods, the new experience, the new values of the sixties have influenced many people, especially the young, and have come to set the tone and trend of education.

During the summer of 1972, a low point in the movement for 'free' schools and universities, Michael Rossman, one of the new teachers to come out of the American student movement, maintained that the number of new-style educational establishments in the States would increase steadily. The number of teachers and parents who believe in the new methods – and among whom an ever-greater number are former pupils of new-style establishments – was on the increase by a measurable curve. In 1973 there were 1,600 progressive schools with 50,000 pupils; in 1975, Rossman calculated, there would be 6,500 schools with 200,000, and in 1977 25,000 schools with 800,000 pupils.

This optimistic reformer has described the spread of counter-culture in the US as an 'alternative system' which is largely decentralized. We must look on it as a network whose constituents are varied and flexible. The net links many different groups, each of which has been established in order to learn new information or to begin the necessary action to make this information available: free clinics, free schools, underground newspapers and radio stations, crisis centres, free universities, student-run experimental colleges, media collectives, anti-war and resistance groups, centres for 'human growth' (through meditation, Yoga, Aikido), environmental and consumer initiatives.

The future begins in the schools

During the past ten years I have studied a bewildering multiplicity of new educational projects both personally and through specialist literature. Several main trends are recognizable:
◇   from guidance to self-discovery;
◇   from obedient receptivity to critical judgment;
◇   from learning existing facts and skills to discovering new ones;
◇   from isolation to openness.
All these tendencies assist and strengthen one another. Although they rarely occur all together in the same experiment, they belong to one single great movement which already enables us to recognize what kind of person the adult of the future will be.

Children at the Peninsula School in San Francisco teach one another; eight- to twelve-year-olds in Shanghai practise learning through teaching; older children in 'open classes' at schools in Hamburg and Leicester are no longer dragooned on their benches for hourly handouts of knowledge, but move from room to room wherever a project attracts their interest. These children will be the thirty-five-year-olds of the year 2000. The self-confident pupils of the famous Oslo experimental school, the apprentices in East Berlin who, as 'masters of tomorrow', are building a new calculating machine almost without adult help, the English schoolchildren working according to the Nuffield Foundation's method of 'learning through research', absorbing natural sciences by remaking the discoveries of the past and treading themselves the difficult path of error and inspiration – all these will hardly allow themselves to be moulded into the smooth working of the 'technocratic society' as conceived by certain industrial futurologists in their models for an authoritarian state.

People educated by these methods differ from their predecessors in that they see the vast gap between what is and what could be; pupils of the creative 'open' methods of Torrance, Flescher, Landau, Ammon, or Wollenschläger already understand that they will have to live in a closed, anti-creative social reality. The conflict they experience in their daily struggle with the adult world will be intensified when they leave school. Young people who trusted the ideas in their school books and fought against a disappointing world used to be told: 'You'll soon settle down.' In the future, this prognosis will no longer apply universally. Even today it is no longer quite correct: readiness to fit in is gradually going out of fashion. No longer is the longing to make the world a better place sloughed off with the first job as a youthful foolishness; fewer and fewer people of twenty and thirty are prepared to toe the line uncritically.

The new teaching prepares its pupils for the difficulties and temptations that lie ahead. And in the spirit of 'learning by doing it yourself' the young are often beginning, even while still at school, to put into practice their confrontation with their rulers. The 'difficult' children who rebel in Germany, France, England, Scandinavia, China and the US are not showing signs of decadence but of nascent maturity: they are beginning to recognize deception and neglect. 'We learn not for school but for life' – that slogan is at last being taken seriously. Life is not a place for false peace and quiet, but for continuous confrontation.

Schools like this are no longer factories for making citizen subjects or obedient consumers like those described by Ivan Illich in *Deschooling Society* (New York, 1971): they are seed-beds of social change. They cannot themselves bring about this change, but they can produce self-confident, creative human beings who will work for it. For them the future will not be an endless series of interlocking prisons or the scene of inevitable catastrophes, but the foundation for a more humane way of life and of living together.

Can foresight be taught?

One Monday morning pupils of Melbourne High School, a few miles from the American space station at Cape Canaveral, Florida, found the front row of benches roped off. They thought that repairs of some kind were about to be undertaken. They hardly noticed the change and simply moved a little closer together. On Tuesday the second and third rows were blocked off, and on Wednesday the fourth, fifth and sixth as well. It was now getting difficult to find room for everyone. On Thursday half the room was off limits, and on Friday they finally had to stand closely packed in the space between the desks and the wall because all the places were roped off.

That was how a teacher taught his pupils about lack of space through overpopulation. Why had they not asked on the first day what was the matter? Why did they do nothing on the second, third and fourth days? Why had not a single one of them dared to

cut through the ropes or climb over? This was the theme of the
heated debates that followed the event. It was an unusual beginning
for the 'Course for the Twenty-first Century' which the futurologist
Alvin Toffler had thought up for his pupils.

Toffler has always been interested in youth problems. He invented
the profession of 'value forecaster', a person who uses changing
values, especially among different generations, as signs for future
ideological development. A university teacher as well as a publicist,
he devoted himself to change in education. Here he found himself
on the same wavelength as Jerome Bruner of Harvard and Oxford,
who considered the uncertainty about the future among young
people to be the most important problem in his work; and also as
Kenneth Kenniston of Yale, who said that the heaviest burden the
younger generation had to bear was 'loss of a vision of the future'.

These three personalities are chiefly responsible for the flood of
futurological lectures, seminars, courses and school exercises that has
been sweeping Canada and the US in recent years. In 1968 there
were fourteen regular courses in the US on questions of the future.
According to Bill Rojas, a young teacher specializing in this field,
there were over a hundred in 1971 and four hundred in 1972.

Here are a few of the methods that have been tried:

◊   In Florida students tried to prepare for the 'future shock' by
moving from one family to another at ever shorter intervals. They
began by staying a month; then three weeks, then two, then one.
Finally there was a change of scene every day.

◊   In Minneapolis students were encouraged to have a 'Be Another
Person Day' at least once a week. They tried to enter into another
personality in order to achieve a higher degree of mental elasticity.

◊   At Mark's Meadow School in Amherst, Massachusetts, the
biology teacher Patricia Burke tried to show her pupils in the first to
sixth grades the possibilities and dangers of future biological engineering. She gave them a box of body components and told them
to make a 'dream body'; then they had to give reasons for preferring
their 'new Adam' to the old. The most popular model for the future
was a person with wings.

◊   In Berkeley in 1970, John Dieges and Ed Schlossberg, both
pupils of the well-known architect and philosopher Buckminster
Fuller, got children aged eight to fourteen to build cities of the
future. Each school day represented one year of the future, from
1971 to 2000. The various designs were used as a basis for discussions
on possible changes in future thought and life-styles. All this was
filmed on videotape so that the eighteen- to twenty-four-year-olds

of the year 1980 will be able to check their ideas against the future that will by then have become the present.

◇   Professor Dennis Livingston, formerly of Case Western Reserve University, and Judi Dressel, of Fox School, Belmont, Massachusetts, recommend that science fiction should be given a place in the school curriculum. Discussions arising from this literature deal, for instance, with the differences between natural science and fantasy science. 'What would you do if you suddenly met a monster from another planet on the way home?' This question evokes the children's fears and prejudices about racial minorities and their role in the future, and these can then be analysed.

Some of these experiments are brash, clumsy and time-bound. But that should not lead us to condemn them. As the famous anthropologist Margaret Mead once elaborated, there have always been hundreds of teachers teaching us about the past, but there are still far too few who discuss with their pupils the subject that must interest them above all others: the prospects for tomorrow's world. And we have far too few independent futurological research stations where critical, systematic and even unconventional research can be done on ways into the future without the need to kowtow to any sponsor. The prognoses supported by industry or the state are still largely a modern sequel to oracles and court astrology.

The desired goal of making people more open, more flexible and more imaginative can be reached not only by discovering and inventing new ways of living, but also by learning to judge and observe more clearly and with less prejudice, and by uncovering long-buried types of sensitivity. Heinz von Foerster and his colleagues at the University of Illinois have long been trying to revive man's trapped or atrophied powers of perception and recognition:

> When from time to time we look through the windows of our laboratory into the affairs of the world, we become more and more distressed by what we now observe. The world seems to be in the grip of a fast-spreading disease which, by now, has assumed almost global dimensions. In the individual the symptoms of the disorder manifest themselves by a progressive corruption of this faculty to perceive. . . . We seem to be brought up in a world seen through descriptions by others rather than through our own perceptions.

As an example for this claim Foerster adduces an experiment devised by George Miller: adults and children were asked to arrange in order thirty-six cards, each with one word printed on it. All the adults wanted to group the words according to their grammatical characteristics, dividing them neatly into nouns, adjectives, verbs and conjunctions. The children – again without exception – paid no attention whatever to this artificial syntactical order, but arranged them according to experiences and perceptions. The verb 'to eat' naturally comes next to 'apple'; 'air' is 'cold'; the 'foot' is used 'to jump'; people 'live' in a 'house'; 'sugar' is 'sweet'; and the combination of 'doctor', 'needle', 'to suffer', 'to cry' and 'sad' tells a whole story.

But modern education, according to Foerster, usually destroys this meaningful way of regarding the world. Most people do not see because they do not wish to see, but prefer to pursue their inbred stereotyped conceptions; and they do not hear, because they only want to hear what agrees with their own previously determined opinions. The 'silent majority' is really a deaf and blind majority that refuses to take in anything that does not fit into its accustomed notions.

These observations are important for the criticism of the present educational system. Examples which were once true but have been overtaken by time are hammered into children's heads by teachers who ask them for the 'correct' answer – i.e., for the preconceived answer. As long as this continues, it will be impossible to understand either the present or the future. For the future will always be different.

Teachers who expect their pupils to repeat what has long been known and accepted as correct should be superseded by a new breed who would ask questions to which the answers are still unknown. Instead of calling on the pupil's memory, they would call on his unprejudiced and unbiased powers of observation and imagination. They would be the opposite of Dostoyevsky's Grand Inquisitor, who wanted to burn the unexpectedly risen Christ as a disturber of the peace, in order to maintain the existing order.

Foerster complains of the blind and deaf products of the present school system not only that they are ignorant because they cannot see, hear or feel, but that they do not even want to be any different. Like many others whose criticism is justified, at this point he too falls into cliché. The truth, as we have seen in several examples, is that especially in the world of school-teaching many people are learning to comprehend reality critically and with open minds and to realize that it is a changing process which they themselves can further and influence.

Schools without walls

The distant green star begins to grow until the whole horizon is filled with its radiance. Suddenly it bursts into shining splinters and threads which immediately close up again to form an orange-yellow ball of light. Red flames shoot out from this new sun; they group themselves into spheres which move apart and come together to the sound of a singing voice.

This cosmic spectacle is astonishing in its continual creation, decline and rebirth. It is projected on to the circular glass screen of an oscilloscope standing in a darkened room of the mausoleum-like Palace of Fine Arts in San Francisco. These instruments are normally used for physical and chemical measurements. But in this case they are dispensing pleasure and a little information. The name of the toy is 'vidium'. It is connected to a microphone: visitors to the exhibition can speak, sing or call into it, and the different sound and pitch of their voices produces different patterns in different colours. Children especially can never tear themselves away.

This device is one of the many exhibits in the 'exploratorium' that was set up in 1968. According to the founder and director of this unusual museum, visitors are supposed to realize that scientifically explicable processes can be conveyed not only by factual description, but also through experiences of great beauty.

The exhibition is housed in an old hall, previously used as an army garage, which has now become a place of wonder and amazement. Its founder bears a famous name: he is called Frank Oppenheimer. My real reason for going to see him was to talk about his brother J. Robert Oppenheimer, the co-developer of the atom bomb. But we hardly got around to that subject. Frank, whose life has been passed in the shadow of his elder brother, preferred to talk of his own plans rather than about the past. 'Oppie' had always seemed remote and supercilious; Frank, on the other hand, captivated me by his warmth, directness and simplicity.

As a pacifist he was for years politically suspect, persecuted and prevented from doing his chosen job. He took refuge on the family ranch. Then, when the witch-hunt ran out of steam, he went back to physics teaching and, with the aid of funds from foundations, built this new 'School for Everyman'. 'What I want,' he told me at his house on a hilltop in the great harbour town, 'is to make the visitors feel what moves the real researcher. Not only the desire for knowledge, fame and power, but aesthetic feelings too, and the pleasure of fantastic ideas and daring conjectures. Science has absolutely no need to be as cold and devoid of feeling, as abstract and dry as it is made out in school books and ordinary teaching. It can move anyone – adults and children alike – just as nature moves them to understand it.'

Oppenheimer's 'exploratorium' has about 300,000 visitors a year. The workshop is open to everyone, and everyone can see how the exhibits are put together; everyone is allowed to play with the 'crazy' machines, even if they sometimes get broken in the process: the magic tree whose hundreds of lights flash on when someone speaks to it; the reflection system that multiplies coloured lights a thousandfold; the bicycle generator on which everyone can make his own energy and light; the perspective room, in one part of which the visitor looks too large while in another he looks too small. Or they can admire the multicoloured sun pictures thrown on a glass screen by prisms or reflectors. A notice board says: 'It only works when the sun shines. . . . Please come back another time.'

There is no research fanaticism, no esoteric seclusion, no secrecy, no VIP fuss; the style here is quite different. The younger brother of the symbol of all brilliant major scientists works as a modest, easily accessible teacher serving Everyman, young and old, and trying to delight him as well as to impart knowledge.

This changed attitude among teachers (with or without diplomas) heralds a new era. The focal point is not the teacher, but the learner who wants to gain knowledge and skill. But he is no longer imprisoned in a school that shuts out the world and follows curricula which are nearly always behind the reality which is developing much faster and more diversely. School classes are beginning to study the mass media and their contents, their methods of representation, their conscious or unconscious manipulation, and what they can and cannot say or can only suggest. A teacher like John Bremer, the founder of the 'School without Walls' in Philadelphia, takes the whole city for his classroom. The pupils study in hospitals, multi-storey car parks, sewer systems, power plants, cattle markets, abattoirs, slums and affluent residential areas, union headquarters and office blocks. They are even admitted to the 'no-man's land' of big business, and they soon notice whether they are being fobbed off with the public relations material or told the truth. For, unlike their parents, they have learnt how to observe, how to ask searching questions, how to form their own ideas. They know all the tricks of advertising, they can analyse and see through posters, television spots and slogans; they even know who holds the power and can read between the lines of economic reports.

Education no longer needs to be training in naïve credulity and servility. The era of cramped horizons, of enforced apathy, and of the lifelong, more or less resigned drift known as a career could be coming to an end if the number and influence of the enlighteners in many countries continues to increase – teachers like Ivan Illich, George Dennison, George Richmond, Anne Long, John Holt, Herbert Kohl, Jonathan Kozol, Patrick Zimmermann, Pierre Bertaux, André Mahé, Jules Celma, Francine Dubreucq, Ernst Jouhy, Dieter Baacke, Hans Giffhorn, Diethard Kerbs and Hartmut von Hentig, to name but a few. They all hold different views, but their emancipatory goals are similar.

'We want to help you so that you will never be taken for a ride, cheated or done down,' says the new textbook for social and community studies published in West Germany by Georg Fischer and his collaborators; 'and we hope that you will learn in good time how to defend yourselves, and to stand up for your rights.'

Against the new illiteracy

A few years ago, the new Paracelsus Hospital in the West German industrial town of Marl was being inaugurated with the usual ceremony. Suddenly, in place of the opening speech, a heated discussion came across the amplifier:

'We'll never be able to afford a thing like that.'

'Our town simply doesn't have the means for a modern hospital.'

'Well, let's give it a try.'

These were excerpts from tapes cut years before at a debate between the town councillors. Against the backdrop of these discouraging remarks the building – complete in spite of every difficulty – appeared all the more impressive.

The tapes had been made by a group of Marl citizens who used to meet at a local recreation centre called 'The Island'. They decided to enrich the adult education curriculum with some of their own data reports. This collecting of information on a large number of local problems and events taught them much more, and much more intensively, about national economics, community problems, and the needs, complaints and hopes of their fellow-citizens than any lecture or seminar could have done.

The Brazilian educationalist Paolo Freire introduced similar methods of teaching into adult education. These methods demanded active participation and tried to awaken a greater degree of receptivity to the information offered by arousing the personal interest of the students. Freire was born in Recife in 1921, the son of middle-class parents. In the depression his parents lost their modest fortune, and so he became familiar with hunger from his early youth. After a few years as a lawyer he turned to teaching because he said he was sick of defending the rights of those with enough to eat.

So Freire became one of the collaborators in the Everyman Project. His friend and fellow-fighter Ivan Illich says that Freire's work shows that every 'awakened' person can learn to read and write in a few weeks. Freire sends his lay teachers into a village in order to discover the political key words, which vary from place to place and from year to year. It may be the spring to which the landowner denies access, or the payment exacted by the police. In the evening, Illich reports, the teacher brings together the people who are concerned about these words, and lets them talk. When a key word comes up he points to a board on which the word has been written. Writing means taking alienated reality into one's hands. After a few hours the adults recognize a dozen key words, and before long they can build up their whole vocabulary out of the syllables of these controversial words.

Since 1964 Freire has lived as a political refugee abroad. For him, reading and writing are merely instruments in the process of *conscientizaçao* – becoming conscious of human and social relationships. 'Conscientization' is far more than the teaching of a skill: it is meant to make the student conscious of his social and historical role. In Freire's words, we have gradually to take history into our hands and mould it instead of letting it mould us.

Today Freire works for the World Ecumenical Council in Geneva. His concept can be applied just as well in the developed countries, where the old form of illiteracy has been replaced by a new variant of non-comprehension. The inhabitants of the highly industrialized countries can read and write, of course, but most of them lack any knowledge of the complex structure of economic interests with which their life and work are bound up. They hardly understand the background to their dependence. They do not know why they do the work they are able to find, nor why they have to consume what they are offered.

But it has been shown that illiteracy in the field of economics, science and technology cannot be eradicated by the usual teaching methods. The interest in adult education all over the world is still much too limited, and the beneficiaries of efforts at 'permanent education' are very irregular in their attendance and often tired out from work: it is obvious that new methods are required, a style of teaching and learning where the student is no longer – to use Freire's words – 'a passive vessel waiting to be filled by the teacher'.

Freire is a forerunner and a present exception. The vast majority of institutions for adult education and vocational training have hardly been touched by the ideas and experience of the new teaching. Often they continue to treat the student as a mentally dependent listener and culture-consumer. He is rarely trained to use his critical faculties or creative powers. Freire defines most teachers as 'bankers with knowledge accounts': they invest facts in underdeveloped people who will later pay them back with compound interest.

But that state of affairs cannot and will not continue, first of all because nearly everywhere adult education has become the centre of public interest. International organizations such as UNESCO, the Council of Europe and the OECD have declared permanent education to be the chief concern for the next decade. The sharp division between school and the rest of life is to be eliminated. The continuity of learning which Jerome Bruner of Harvard and Oxford considers one of the most important developments of the future will be reflected in a profound change in the life-pattern of every individual.

Plans are already well advanced, and when they are put into practice a five-year-old, for instance, will have roughly seventeen years of education and thirty-seven years of work before him until his sixtieth birthday. He will enter upon his working life at the age of twenty, and from then on he will be able to expect approximately six months of paid education in every five years.

According to a study by Dieter Mertens of the German Federal Institute for Employment in Nuremberg, by 1985 a typical educational curriculum vitae in the OECD countries would look like this:

Age 20: working life begins.

Age 25: first stage of further education, which can take the form of a vocational 'topping-up of knowledge'.

Age 29: second six-month phase of further education. At this stage it is less a question of acquiring specialized knowledge than of gaining 'key qualifications' intended to broaden the personality: qualities such as the ability to go on absorbing permanent education, to acquire and use information, to understand and change one's role in society, to see one's practical work in perspective by theoreticizing it, to link theory with practical qualities, to develop one's creativity and technical insight; the power to analyse the interests of different groups or classes; the ability to plan, to communicate, to continue learning, to make the best use of time and of the tools of one's trade or profession, to establish goals, to collaborate with others. In this phase it is intended to show the student ways towards concentration, accuracy, the reasonable solution of conflicts, the sharing of responsibility and the pleasure of achievement.

Age 33: third stage of further education. This is mainly to teach new elements in education which did not form part of the curriculum at the time of the student's own schooling. It would prevent people from being overtaken by their juniors.

Age 37: fourth stage of further education. This could be mainly for specialized professional training.

Age 41: fifth stage of further education. Here the student could concentrate on freely chosen subjects with a wider background, not necessarily connected with his own profession.

Age 45: sixth stage. The main aim here is to secure the student's professional status.

Mertens intends this curriculum vitae to be no more than a model for discussion. He thinks that quite different patterns might be preferred, and sets out the following alternative possibility: between the ages of 22 and 30, everyone would have the right to four weeks' further education a year; between 28 and 32 a whole year could be taken; between 33 and 39 there would be an eight-week educational period in each year; and between 40 and 50 six months would be reserved for further education.

There is much to be said for and against this model. On the one hand it seems too heavily weighted in favour of vocational training, putting too little emphasis on social tasks and faculties not used professionally. Industry, on the other hand, criticizes the high cost and the loss of working hours. The debates over the French *éducation permanente*, introduced in 1970, and over the West German unions' demand for educational leave, gave a foretaste of the kind of confrontation we can expect. These are social conflicts on ground that has never been fought over before.

It is significant for the future that permanent education is everywhere regarded as a central social must. A new and enormously important sector of public life is opening up. If we accept that in the near future the discovery and development of man's resources will become as important as the discovery and use of natural resources has been for the last two centuries, then adult education is the sector where these efforts will be most strongly developed.

But to assume that official institutions with their almost inevitably prescribed and therefore inflexible character will have a sort of monopoly in the field is to have too little faith in the people of the future. They will doubtless develop many different varieties of 'lifelong learning'. And if they succeed in realizing the suggestion whereby every adult receives vouchers to enable him to spend his educational credit as he wishes in the periods allotted for it, then a multitude of novel, unconventional and experimental adult colleges, study clubs, free seminars, summer courses, travel groups and research communities will flourish.

Illich and his followers are no longer alone in advocating such generous arrangements. If the movement should become institutionalized, the official institutions would still be supplemented by groups of adults cultivating and developing the new outlook and method. Trust in people instead of control, an increasing capacity for thinking and planning, participation and self-government in as many fields as possible by a growing number of people, are all expressions of a new type of motivation. Even countries with totalitarian or centralized governments will not be able to avoid it. In the long run they will no more be able to resist this progressive human revolution than previous generations were able to resist the general urge to abolish slavery.

Taming educational technology

It is a fact that all political systems regard mass education as a necessary achievement; this must mean that much greater resources will be made available, but also that there will be more emphasis on educational control.

The introduction of technical educational aids may provide opportunities for such control. This is not true in the case of substandard film cameras, video recorders, typewriters or film projectors: teachers and students alike can handle them as individually as they can pencils and pens. But expensively produced films, sets of slides, computer teaching programmes, courses on videotape and television broadcasts present another case. These things are produced centrally and usually far from the individual institutions that will use them. Even today we find private or state-controlled 'education industries' beginning to supersede the handicraft phase in education; their importance is bound to grow with the increased use of mechanical teaching aids in schools, universities and adult and vocational education colleges.

One can object that school books have always been produced

centrally and that most curricula are not drawn up in individual schools; but these objections do not eliminate the danger. The suggestive power of the new technical media is incomparably greater, and their cost is sometimes so high that it is bound to reinforce economic concentration in the education industry. But the main purpose of technical aids – at any rate in the minds of many of the people in the education industry – is to save on staff. They maintain that mechanical means of multiplying education have become essential because the supply of teachers is already falling short of the demand and will do so even more in the near future. A comparison is often made with the telephone: if automatic dialling had not been invented in time, the whole system would have broken down long ago through lack of switchboard staff. The entire female population of the US would not be sufficient today to provide telephone operators for the old system.

But we urgently need a technology assessment for the new media that are being introduced into education. It might effectively counter new concentrations of power which could conceivably politicize education in the name of rationalization, with resultant dismissals and prohibitions from teaching.

American plans for an education satellite project are already well advanced and scheduled to be put into practice by the late seventies. They show clearly how much power the education programmers of the future will derive from technical aids. ATS-6 (Applications Technology Satellite No. 6) is to send out the same school programmes to eight American states – Arizona, Colorado, Idaho, Montana, Nevada, New Mexico, Utah and Wyoming – an area of roughly 580,000 square miles with $8\frac{1}{2}$ million inhabitants. Originally there were plans to build many new schools and employ new teachers for this region, but now these plans have been curtailed in accordance with the new project. Only a fifth of the lessons transmitted will be 'home-made' – the rest are to be supplied from an educational deep-freeze. The fact that the system will have two-way capacity is emphasized as a special advantage. It will even 'permit the observation of the children's facial expressions as they receive the programmes'.

Education satellites of this kind could be particularly important for the intellectual and cultural future of the Third World, for they provide a cheaper alternative to building schools and training teachers. Will these countries develop their own television teaching programmes? Or will the new stars in their firmament mainly transmit messages originating in Western or Eastern industrial countries? Already, as Professor Herbert Schiller of the University of North Carolina has shown, American thrillers dominate the entertainment programmes of many developing countries. Is this to be repeated in the schools sector?

If the new technical aids were used for completely different programmes directly controlled by teachers and students, then quicker and more impressive results could be achieved by this method than by the written word, because:

◇   film-making trains the eye and – because it is in the nature of a project – develops team spirit as well;

◇   the logical faculty is developed by playing with computers and 'guided tortoises', as practised by Bob Albrecht at the universally accessible People's Computer Theater at Menlo Park, California, and by Professor S. Papert (MIT) with children between the ages of four and ten in Cambridge;

◇   learning from multiple-choice programmes such as the Ticcit system used at the University of Texas and the PLATO programme at the University of Illinois allows each student not only to progress at his own speed, but also to ask critical questions;

◇   with the help of typewriters and duplicating machines, teachers and pupils can become publishers on a small scale.

Television schools and universities seem particularly promising for the broadening of adult education. The showpiece is the British Open University, which is already demonstrating the teething troubles of permanent education on the widest scale.

Lessons of the Open University

Clearly . . . there can be no compulsion to read this report or to act upon it. For this reason it would be unhelpful to make it the subject of examinations and assessments. Instead, it is offered as a stimulus to action inside and outside the limits of the course of which it is a part. So if you attempt to act on it, do so *for the sake of all things* not for any personal advantage. To attempt this could be a joy, and a revelation. To leave it alone: no blame. What examinations there are will not depend on how you act in response to this item: only on your ability to express your own doubts and certainties about design.

This relates to a course on design and was written by Professor J. Christopher Jones, then of the Open University, in 1972. He was one of the most unusual teachers in Britain's newest and largest

university, which provides tens of thousands of people with a
university education by means of radio and television programmes,
detailed correspondence courses, tutorial classes and summer schools.
He considers the knowing and acquiring of facts to be only a small
proportion of man's abilities: inspiration, feeling, tactile and formal
understanding are far more important.

The incident which brought Jones to my attention was characteristic of him. It was at a congress on technological forecasting held in
Glasgow. A famous Californian expert had just finished lecturing
with clockwork eloquence on his new traffic system, which was
supposed to function with the accuracy of a sophisticated precision
watch. A red-haired man of medium height went up to him and
asked in a voice shaking with emotion:

'Sir, have you read *Mein Kampf*?'

Intimidated by the man's evident rage, the lecturer replied: 'I – er –
don't think so.'

Whereupon the red-haired man flung at him: 'Well, your ideas
sound alarmingly similar: total subjection of human beings!'

That was JCJ all over. He is a Welshman to whom research and
teaching mean not merely discovering and passing on facts, but a
passionate striving after improvement and change. Like all the
prototypes of millennial man, he tried many things before finally
settling down to academic teaching. That is probably why he is such
a good teacher, one who attracts and stimulates his students. It was
no surprise that Chris Jones gave up old-style universities; after years
of successful work at the Manchester Polytechnic, he turned towards the experimental Open University. For here, on the initiative
of the Labour government, a new type of mass experiment in adult
education was taking shape. It immediately attracted hundreds of
academic teachers eager for something new, and weary, like Jones,
of the normal university business.

The percentage of the English population that showed itself interested in this experiment turned out to be smaller than expected.
Every one of Britain's thirty million television owners was given
the chance of three years' study for a degree that would be equivalent
to those conferred by other universities, but 'only' thirty thousand
applied – a sizeable number, but all the same a small percentage of
the possible total. Twenty-five thousand were admitted, and well
over half stayed the course. These are figures that sound impressive
to educational economists. It is obvious that a graduate of the Open
University costs the state far less than one educated at a conventional
place of higher education. But educational psychologists, theorists

and teachers are not satisfied. They think that old wine is simply being poured into new skins – a view that has been expressed particularly by the pioneer of American futurology and communications science, John McHale. 'A new technique is being introduced into a system which is largely obsolete, thereby perpetuating its faults by constant repetition and possibly even by using more bureaucratic methods than before.'

This criticism appears more than justified. In the 'television university', televisual lectures take second place. The main emphasis is on study packs, sent out from a central office in London, which contain material for at least twenty hours of homework a week. Furthermore, the students have to attend regular evening classes conducted by counsellors, many of whom are unsuitable as teachers. There are examinations to be written and questionnaires to be completed. The most interesting feature is the annual two-week summer courses where the students can have discussions in a setting far removed from their work background.

Chris L. Crickmay, a young staff member of the Open University, would like to see more use made of the new and peculiar possibilities of electronic media for teaching. He says, most convincingly, that modern man in his work situation has become a cog in a huge machine which he cannot comprehend. A truly open university would not push him further down the path of specialization, but should help him at last to understand the working of the whole. Instead of opening up vistas, this centrally directed educational system only narrows a man's horizon to certain predetermined social roles and to his job where he cannot act according to his own judgment but has to follow directions.

His programme for an open university in the spirit of emancipated education is summed up as follows:

electric media (not print) dominant, goal-free courses; no plans; decentral control; no committees; no predictions of how people will react; integrative and non-specialist courses; wide and fast channels of communication between everyone involved; no form filling; no exams; local and global needs satisfied; wide responses expected of all concerned (i.e. acting as persons not as roles); trust rather than paternalism; access to knowledge for everyone, not just for members of the institution. If these are desirable shifts, can they be made to happen either from inside or from outside the present Open University, or will it be left for future Open Universities to embody them?

At the moment there is hardly a sign of the Open University being remodelled along these lines. It continues to be used as a rung on the career ladder. It does not fulfil the hope that was the chief impetus at its foundation: so far, the working classes have not paid much attention to the new university. In the first years more than a third of the places were reserved for the educationally under-privileged, but only 6 per cent were actually taken up by such students. Almost 40 per cent of the students are teachers who want to continue their education.

The reason why the original idea did not work as hoped has been carefully examined. The result of these investigations could have been foreseen, even if it had not been so carefully documented: the courses were too academic. The students were not sufficiently pre-pared for the kind of disciplined study that was required, and they found the work hard to understand. Apart from that, the nervous and physical strain had been underestimated. After a long day's work, people are just about able to take in popular television programmes for relaxation. Finally, very few workers were able to find the time for evening classes with tutors or for summer courses. The fees and costs of extras are still too high, especially for the younger workers.

So the Open University is not as open as its name promises. It largely disregards the findings of modern education, the stress of production, the lack of free time, holidays and money. The lesson to be learnt from this large-scale experiment is that permanent education will become more than a slogan to arouse false hopes only when the work-pattern and life-style of the masses are fundamentally changed.

All the same, profound changes in adult education must be ex-pected before the millennium, even if the structures and systems are not completely transformed by then. Why?

◇　New possibilities always bring new demands and solutions in their wake. The invention of the motor-car, for instance, soon pro-duced profound changes in our environment and way of life: people had to adapt themselves to the new means of transport. In the same way the innovation called permanent education, for which many of the preconditions are still lacking, will soon awaken passionate longings and demands and these will eventually produce – by means of political conflict if necessary – the external circum-stances necessary for the enjoyment of the new opportunities.

◇　In the next few decades the increasing rationalization of work processes – especially those requiring physical strength and little intelligence – will mean that only better educated, responsible and creative people will be able to find work. Working hours will be shorter and productivity higher, so that more time becomes available for education. This development will mean that education can become freer and less cramped by the necessity of saving time.

◇ We are only now beginning to study how people learn or what prevents them from learning, how children's creativity can be developed even before they go to school, how the human personality can be activated and released from routine; until now the mass development of all this potential was not even thought desirable. What was wanted were unimaginative obedient underlings who would cause no difficulties. But now we have begun to experiment intensively in psychology, education, behavioural science and other human sciences, with the result that many of the old restraints will soon disappear. We are beginning to understand the conditions for lifelong creativity, and as we recognize them we are forced to create them. 'Children's co-operatives' as set up in empty shops in Berlin, adventure playgrounds, free schools and open universities are only the very beginnings. Those who have passed through them will clamour more and more energetically for a work situation which does not force them to deny their abilities.

In spite of its faults, the Open University is a step in the right direction which may become the impetus for decisive change. The process has already begun. The English trade unions have followed the example of those in Germany and Italy in demanding annual educational holidays so that the potential users of the new institution will have more time. Belgian workers have gone even further and are claiming one free day a week for education. By way of the international agencies, similar demands are penetrating into every national assembly. It is encouraging to find Professor Henri Janne, the Council of Europe's adviser on adult education, proclaiming visions of the future 'which appear revolutionary in the light of western educational tradition':

◇ Rhythm and length of education will no longer depend on rigid prescriptions.

◇ The stages in education will no longer be linked to the student's age. Forms or grades are out of date in so far as they depend on a strict and all-inclusive order of education. Pupils will no longer be grouped according to age: the groups will vary according to the subject and the stage reached.

◇ Degrees will lose their absolute value.

◇   Teachers will no longer be lecturers, but advisers who activate the lessons.

◇   With the help of the teacher, the pupils will plan their own studies. Mechanical teaching aids will be increasingly used.

◇   Study will no longer go on in a place called 'school'. It will be spread over orientation, documentation and information centres, mass media, individual counselling and group work.

◇   The content and methods of adult education can no longer be dictated from above. They will be democratically determined in the light of continuing analyses of educational needs.

The equality factory

The incipient development of hitherto hidden, uncultivated or buried abilities in an ever-increasing circle of personalities ought to be greeted with shouts of joy like the discovery of a constantly re-newed treasure. But for the moment the voices of doubt and fear remain dominant. Sceptics warn us not to overestimate human potential. They maintain that the distribution of ability is hereditary and therefore unequal. Neither new educational methods nor social changes will be able to alter that fact. 'And what are we to do with so many educated, creative people? Who is going to do the boring and dirty work?'

I received the first answer to this question in a medium-sized Swiss firm. Felix Schwarz, the architect who designed this re-frigerator factory, told me that the two owners were 'extraordinary types'. They had demanded that the buildings be so designed that the workers could see both the raw materials arriving and the finished products leaving the factory. In this way they would be able to form a concrete idea of their achievement.

A visit soon showed that this was by no means the only experiment in reform. Every employee, from the highly qualified to the un-skilled, received the same wage; and this was 10 per cent above the top level in comparable Swiss enterprises. This was made possible by not making profit the first consideration, by savings in the accounting system, and by reducing the turnover of labour.

While other concerns often found it difficult to keep their workers, here men voluntarily entered into long contracts. The reasons were to be found not only in the wages, but also in the right of participation and in the variety of work – both factors still rarely found in other firms. One of the practices was for workers to be employed not only on their own special jobs, but also occasionally on others, of both higher and lower status. At first it had not been easy to get this reform accepted. The unskilled workers were reluctant to touch the work of the specialists, the qualified workers felt themselves to be above the labouring jobs. But gradually it appeared that the frequent change of workplace produced satisfaction and emotional bonuses which could not be expressed in figures, but were nevertheless to be felt in the climate of the place. The men found variety in having demands made alternately on their intelligence, their manual dexterity and their physical strength. Even unpleasant and uninteresting jobs are more bearable if they have only to be done occasionally.

At an international congress of IG Metall, Johan Galtung described the following vision of breaking up the rigid work structure:

> What I mean is vertical rotation – for one, two, three days in the week the director is put on the assembly line and the worker is sent to the head office, until they all have roughly the same experience in all parts of the factory. One can expect that this would result very quickly in innovations, which could then lead to a second phase: the general broadening of work, with the consequence that every workplace in the factory would have approximately the same value in terms of demands and stimulation. . . . And why should it be unacceptable that workers, in addition to their usual tasks, should have direct contact with customers and learn of the reaction to their products? Why is it so incomprehensible that office staff should participate more closely in production? And – to go one step further – that customers could be expected to take part in the production process?

Many trade unionists shake their heads at such rash suggestions; but I think that Galtung, the Peer Gynt of sociology, is very realistic, because he takes into account the fact that man's abilities are beginning to be liberated. To hasten the process he has suggested that the unions should take over a few factories and experiment with revitalizing the work. They should even subsidize production in the critical transition period. Changes of this kind in a few factories would then act as catalysts on others.

If the Everyman Project gains in breadth and intensity, one may foresee the time when boredom, monotony and routine will no longer dominate in the world of work. Then everyone, according to his skills and knowledge, could be a teacher as well as a student. In this way there would be a lively exchange of knowledge among people working in completely different fields. A mathematician, for instance, could learn animal husbandry and vegetable-growing from a farmer who would be learning book-keeping and the elements of agricultural surveying from him. When all those who want to learn cease to be kept in dependence and are allowed instead to develop and communicate, then education will no longer be synonymous with a crippling of man's faculties.

A new type of active and permanent education will liberate the powers of criticism, imagination and independence in millions of Everymen, and this will profoundly change our daily life. In the new society the important questions will be: 'Is my work interesting?', 'Is my job satisfying?' or 'Has my daily activity a purpose?' instead of the anxieties which are, in fact, necessary under present conditions, such as: 'How much do I earn?', 'How much power, how much status have I?'

But longings and visions alone do not make reality. That will have to be fought for in the field of politics through fresh, vital and democratic initiatives which will help millennial man to test and develop his growing potential.

3 Technology Tamed

New visions of the future

We came into bright sunshine from the darkness almost without transition. A moment earlier we had been driving at snail's pace through a 'typically English' fog almost impenetrable to the head-lights, and now the driver was dazzled by the sudden sunlight and had to pull in to the side of the road. We climbed out of the bus in the shabby street of a Manchester suburb and began to romp about like a noisy pack of children .The Ferranti computer works where we had been making a television film that morning lay behind us in the greyish-yellow pea-souper; and here, not a mile away, we found ourselves under the clear blue sky of early summer.

That is how I experienced the transition from a polluted to a 'smokeless' zone in the capital of the industrial north-west of England. It was like crossing the frontier between two worlds or two ages. At the time – the early sixties – the Clean Air Act was still quite recent. Great Britain was the first great power to take steps against the disagreeable side-effects of growth by passing an Act of Parliament in 1956. The British are political pragmatists, and they knew that it would be impossible to implement the severe controls that were to be imposed on industrial and private users at one stroke, so they began by creating 'smokeless zones' here and there all over the country. They were to be examples and promises for other areas still steeped in smoke and stench. Fifteen years after this pioneering piece of legislation, a BBC camera team flew over the whole island and noted that 'now, flying over the British countryside on a clear day, a traveller will be struck by the absence of tell-tale smoke plumes from most industrial areas'.

When the British began to clean up their rivers they achieved similar notable, if not total, successes. In 1957 the authorities com-missioned environmental scientists to test the Thames for pollution: on the twenty-five-mile stretch from Richmond to Gravesend they found not a single live fish. Fifteen years later, fifty different varieties were swimming in the same waters. Now they are gradually banishing factories from the shores and turning them into parkland with riverside walks. Industrialization had turned the green valleys of Wales into sterile slopes and slag-heaps; now they are being successfully rehabilitated. Already new recreation zones have been created. In the nineteenth century, machines were euphemistically described as 'iron angels': they set out on their march to conquer the world from the north-west of England. Now this area is being systematically reafforested. By the year 2000, it is hoped that many areas of Britain will be as green as they were in Shakespeare's day. In future the only kinds of industry permitted in the rehabilitated areas will be those that do not contaminate nature or create health risks.

People's wish-fulfilment dreams about their future environment are changing. At the beginning of the twentieth century, the smoke banners of factory chimneys were still seen as banners of prosperity: industry, motors and machines were symbols of progress. Now, at the turn of the millennium, our vision of what the future should and could be is changing radically: technology is no longer in the forefront. Its new role is to be subservient, unobtrusive, inconspicuous, and as harmless as possible; we want to hide it away.

Today high-voltage cables cut across the sky: tomorrow, we conceive, the eye will be free to travel unhindered to the edge of the horizon. No smoke will pollute the air, no waste pollute the water. Noise and depressing, monotonous industrial buildings will be banished: ugliness is on the retreat. Man will escape from the rhythm of the machine: he will be able once more to move at his own pace, to work according to his own sense of time. He will have freed himself from the visible and the invisible chains with which the age of mechanization had bound him.

Styles of technology

Such a description still sounds Utopian. But it is not only possible, it is a very realistic vision of tomorrow's world. Plans, projects, experiments and occasional concrete examples all point in the new direction. Many things that only yesterday were rejected as being uneconomic and therefore impossible have turned out to be perfectly feasible. For instance, at the beginning of the sixties it was thought that to put cables underground was too expensive and could only be undertaken in exceptional circumstances. By the beginning of the seventies the awakened interest in the environment had resulted in hundreds of miles of cables being laid underground. On the one hand, the work had become less costly owing to improvements in underground construction methods and electrical engineering; on the other, public and private enterprise suddenly found it could afford the extra expense which only recently had been considered out of the question.

An important role in this change of attitude was played by the intellectuals with their criticism of our civilization. At first what they said went unnoticed; then it was derided. But even at a time when the prestige of technology was still at its height and criticism was considered reactionary and heretical, these warnings made people feel uneasy about the machines that were beginning to threaten life and often to destroy it irrevocably. But soon it turned out that the intellectuals in their supposed ivory towers had been able to arrive at a more correct judgment of reality than the so-called pragmatists because their frame of reference was wider. The 'realists', on the other hand, who had refused to see the side-effects or the long-term results of industrialization, appeared much less serious – in spite of their strictly rational attitude – than Charlie Chaplin with his comic genius. The conveyor-belt sequence in his masterpiece *Modern Times* shows how humiliating and ridiculous inhumane production methods can become.

But unease did not change into worldwide terror until the bounds of technological development had been overstepped and the whole of mankind put at risk by the atomic explosions of Hiroshima, Bikini, Eniwetok and Novaya Semlya. This finally shook the dogma of the purity and inevitability of technical progress to which everything and everyone was to be subordinated.

When Anthony Wedgwood Benn, a leading figure in the British Labour Party, was Minister of Technology at the beginning of the seventies, he spoke of 'the central problem of the present day', which is whether men will retain control over the machines they have built, or whether they will let the machines roll over them. The risk is very real, he said, that we shall be exposed to the man-made environment just as cavemen once were to nature. They were surrounded by forces which they did not understand and lived in constant fear; that could happen again.

But can we really speak of 'technology' *tout court*? The English engineer and inventor Professor Meredith Thring, of Queen Mary College, London University, feels that we have not really tried to construct a 'creative technology', but have been content with 'cheap technology' whose only object is to produce inexpensively, profitably, economically and quickly. He thinks that experimental technologists and engineers could have constructed machinery far less hostile to man and his environment if only that had been considered as important as purely economic factors.

The overriding motive in the world of machines is the employers' motto: More output! More profit! But from a more detached viewpoint this formula turns out to be deceptive: it is true that technology aimed purely at efficiency and profit does bring quick rewards, but because of its harmful side-effects it also produces high social costs and long-term losses.

Technology is like architecture: its style reflects the social, economic, intellectual and spiritual values of its age, and they in turn are influenced by its achievements. The Belgian historian of civilization Henri van Lier of Brussels University thinks that the brutal 'dynamic technology' of the nineteenth and twentieth centuries, which aimed at subjugating man, will be followed by a 'dialectic technology' whose beginnings can already be detected. He hopes that this new technology will carry on a dialogue with nature and with man, and that instead of remaining outside the sphere of civilization it will become part of man's cultural achievement, a sister to the arts.

In my research I have uncovered numerous attempts to create a new technology, a technology which will enable people to live with it instead of suffering under or rebelling against its yoke. I should like to group these attempts into four main trends:

◇   The first and most highly developed seeks to establish strict controls over technology.

◇   The second seeks to reduce technology to a minimum.

◇   The third and most imaginative aims to change radically the nature and functioning of technology, to bring it closer to nature, to tame it by 'controlled evolution'.

◇   The fourth trend is to change the direction of technology towards ends that are more beneficial to man and his environment.

Naturally these categories overlap, merge and complement one another. Technology has its controllers, its ascetics, its transformers, its pilots; they will all try, by combined and by parallel efforts, to create a new and more amicable relationship between man and his tools, so that he may survive the crises of this millennium.

Pollution-rush millionaires

Around the end of the sixties and the beginning of the seventies many countries witnessed the development of a billion-dollar industry for the control of environmental damage. The world market was beginning to suffer a stagnation or even a fall in the profits from armaments, and unexpectedly large new profits began to emerge from the manufacture of automatic instruments for detecting and measuring dangerous substances in the air, in water, in the soil, and in foodstuffs; chemical and pharmaceutical products to fight pollution; and air filters, water purification plants and installations for the recycling of waste materials.

It was the beginning of a 'pollution rush' like the gold rush in Alaska. A few particularly skilful people made millions from it in the shortest possible time. One of them was the American Robert L. Chambers, who borrowed money to found the firm of Environtech in Menlo Park, California: after less than three years the firm's annual earnings were more than $150 million, and its income was still rising rapidly.

The United Nations Environment Agency has estimated that by 1985 about a fifth of all industrial products will be for the purpose of reducing or eliminating the undesirable side-effects of technology. The sort of sums that will be involved can be estimated from the Westminster Dredging Group's project for an artificial rubbish-dump island in the North Sea. It is to cost approximately £50 million. It will be about fifty miles away from any human habitation and will process a fifth of Holland's waste, in particular dangerous chemical waste and about 200,000 used cars annually; after processing it will supposedly be safe to dump this waste into the sea. As usual, American statistics are even more colossal: as long ago as 1971 the US was spending $3·5 billion on various methods for disposing of 112 million tons of paper and plastic, 16 million tons of glass and 14 million tons of metal.

Thus enormous new financial burdens are being created. In the first instance, of course, they must be borne by industry on the grounds that industry is responsible, but eventually they will fall on the public in the form of higher prices and taxes. Moreover, the huge companies of the motor, airline, plastics and chemical industries which are mainly to blame for pollution were quick to buy up small companies producing anti-pollution products, or to start new ones themselves, and so they were able to share in the boom and actually make profits from the elimination of the damage they themselves had caused.

A further cause for anxiety is the fact that this new industry for the preservation of the environment actually helps to prey on the environment. Not only does it use up raw materials, but frequently it does environmental harm. The charming picture I drew of the cleansing of the English skies has its darker side: the process depended to a large extent on the elimination of sulphur from domestic coal, with the result that the air near desulphurization plants became more polluted than anywhere else.

Environmental early warning: technology assessment

Facts such as these make one wonder whether it would not be wiser to stop trying to make good the harm done by technology and to concentrate on preventive measures as soon as possible. The destructive side-effects of industrial processes can rarely be eliminated afterwards and ought to be prevented in the first place.

Considerations such as these resulted in a new concept, at first in the US but soon in other highly industrialized countries as well. This concept is known as 'technology assessment', and in the seventies and eighties it will be as controversial as self-determination, citizens' self-help and anti-authoritarian education are today. Technology assessment can be compared to the controls and trials that are necessary before any new drug or medicine can be put on the market. But it is much harder to assess the drug called technology. It is not enough to consider possible health risks; the assessment must extend to the social chain reactions that can proceed from technical innovations – as the television and the motor car have shown (see p. 107).

The father of TA – the international abbreviation for technology assessment – is generally acknowledged to be an American Congressman of many years' standing, Emilio Q. Daddario, the son of Italian immigrants. His career is significant, because he is one of a small group of politicians and statesmen who were trained as lawyers or economists but who have been forced to concern themselves with natural science and technology – subjects originally quite foreign to them – because of the growing importance of these subjects.

During the Second World War Daddario was still training for the law. Because of his knowledge of languages he was assigned to an intelligence department dealing with German weapons research. He became deeply impressed with the importance of applied research and technological development for strategic and political decision-making; and he also realized that his knowledge of these subjects was inadequate for him to play his part as a well-informed citizen. He was elected to Congress in 1959, and from that moment became deeply concerned about this gap in his experience. The US, at that time, was suffering from 'Sputnik shock' because in 1957 the Soviets had launched a satellite round the earth before the Americans had managed to do so. A crash programme was introduced to help American universities, research institutes and laboratories regain their leading position in the world, and Congressmen suddenly found themselves in charge of allotting eight-figure sums for research without having any idea whether the money was being well spent.

'Unlike most of my colleagues,' Daddario told me when I interviewed him during one of his European trips, 'I was worried even then that we were giving so much priority to space research. But at the time it was difficult for me to weigh up the pros and cons. So I was glad when, in 1963, we set up a sub-committee of the Congressional Space Committee to inquire into science, research and development in a wider context.'

Daddario was still considered a newcomer to Congress, but owing to his contacts with the scientific world he soon became one of the best known and most popular Congressmen. One of his most interesting initiatives was the strengthening of the scientific information services for Congress and the improvement of its access to scientific documentation. At last the Representatives had sufficient expert advice to be able to argue with the government and its staff of experts on a proper scientific basis.

That is the history of the developments that finally led to technology assessment. Graham Chedd, an English scientific journalist working in Washington, reported that the idea was born early in 1965 at a table in the restaurant of the House of Representatives. Four men, including Daddario, who had breakfast there every morning, discussed at one of these routine encounters a remark made a few days before by Jerome Wiesner, professor at MIT and former scientific adviser to the President, to the effect that the US needed an early-warning system to protect people from the consequences of their inventions.

It took the influence of yet another man to turn this suggestion (to which Daddario had been very responsive) into a concrete proposal for new legislation. Charles Lindbergh, the famous aeronaut who had made the first solo flight across the Atlantic in 1927, published an article in *Reader's Digest* called 'Is Civilization Progress?'

It was one of the first criticisms of industrial growth to be made in the technology-mad US. Daddario immediately tried to make contact with 'Lucky Lindy'. It turned out to be difficult, because the one-time darling of the nation had lost much of his popularity owing to his period of sympathy with Hitler, and this had embittered him and made him shun public life.

However, he finally received the Congressman, and their conversation completely convinced Daddario that Congress urgently needed an agency for assessing technical innovations and controlling them by law. In 1967 the first draft of a law for the foundation of an 'Office of Technology Assessment' came before the House; but it took almost six years to push through this Bill for a new democratic institution against opposition from the most varied interests.

Turning-point for technology: the supersonic jet

In the debates about TA (translated as 'technology arrestment' by its opponents) the struggle over government finance for supersonic civil aircraft (supersonic transport – SST) played an important part. This was understandable, because the question of encouraging or preventing the introduction of this new type of aircraft, which could fly from New York to Europe in two and a half hours and to Bombay in five, was the perfect example of the kind of problem with which the proposed Office of Technology Assessment would have to deal. It was the first time since the Industrial Revolution – and that is what makes this occasion historically so significant – that people had not dashed blindfold into a new technical adventure, but had been prepared to wait until possible drawbacks had been considered.

Of course it was already well known that aircraft travelling at such speed produced powerful shock waves resulting in a supersonic boom. This had been experienced with the supersonic planes already in military use. The Air Force pilots gaily called it the 'sky fart': the noise was nearly as loud as a bomb explosion, and the vibrations had harmful effects on a wide area. Windows were shattered; houses collapsed; people's sleep was disturbed; pregnant animals miscarried.

As long as these things happened only in the thinly populated areas where military practice flights took place, the general public hardly noticed them. But the new supersonic aircraft were to fly to and over the great cities of the US. The areas near the airports would suffer most, because an SST landing or take-off makes appreciably more noise than that of an ordinary jet. How would the civilian population react?

The champions of the project – Congressmen from states with important aircraft industries as well as lobbyists and government spokesmen – maintained that things would not be so bad: people would just get used to the noise. At the Congressional hearings in May 1960 they managed to reassure the Congressmen so successfully that by August 1961 they were ready to vote the first annual instalment of $11 million for a two-year research project to iron out the last snags in the SSTs. But more and more improvements turned out to be necessary. This was because warnings were gradually beginning to seep across the Atlantic from the Swedish Institute for Aeronautic Research under Dr Bo Lundberg: not only would people and objects under the flight path of the SSTs be affected, but the health of the passengers and pilots might also be in danger. At heights of 65,000 to 88,000 feet they would be exposed to cosmic radiation, but the aircraft would have to fly at such altitudes for most of their journey to reduce air resistance.

The prototypes for the giant bird were already being built in the hangars of the Boeing Corporation near Seattle; but meanwhile a formidable opposition was gathering in the US itself and beginning to alarm the public with well-founded objections to this 'expensive, dangerous and irresponsible prestige project'. The enemies of the SST could not be dismissed as tiresome grumblers, as the public relations staff of the aircraft industry tried to make out: William A. Shurcliff, a nuclear physicist at Harvard, had the reputation of being a reliable, unemotional and scrupulous scientist. In the course of his career he had acquired not only a thorough knowledge in his field, but also the conviction that scientific and technological enterprises should be accompanied by a sense of responsibility for their consequences. During the Second World War he had been right-hand man to Vannevar Bush, the head of the Office of Scientific Research, where new weapons were developed. So from the forties onwards, Shurcliff had been involved in the great crisis of conscience over the making and dropping of the atom bomb. That was the time when many scientists began to feel guilty and to resolve never again to lose sight of the consequences of their work.

In 1945 Shurcliff had shown great care in helping to produce the Smythe Report, the official American report on the atom bomb project. Now, inspired by a letter from a biochemist which appeared in the *New York Times*, he devoted the same care to studying all the facts relevant to the SST project. He soon concluded that once again the men in power and the scientists, with their fatal inclination to pursue technical virtuosity, were collaborating without a thought for the possible harm they might do; and he decided to ward off this threat.

He contacted personalities who had already publicly voiced their doubts about the project, and with them he founded the Citizens' League Against the Sonic Boom. They all devoted part of their savings to full-page advertisements in the most influential newspapers, in which they asked the public to support them in their protest.

I visited Shurcliff in his little wooden family house in a quiet side street in the university town of Cambridge, Massachusetts. His old-fashioned living room overflowed with membership cards, pamphlets and leaflets, and here I realized that the American ideal of grass-roots democracy still flourished in spite of mammoth bureaucratic institutions. From this modest room Shurcliff and his family had waged a tireless battle against the whole establishment: generals, Senators, lobbyists, leading members of the Federal Aviation Administration, industrial bosses and finally even President Nixon himself were forced to modify and then to abandon their plans. In December 1971 Shurcliff and his growing number of helpers achieved their goal: the Senate turned down the new allocation for the SST project. The Boeing Corporation was forced to give up its controversial plans.

Shurcliff, who was no longer young but extraordinarily lively and energetic, described to me how he had achieved his purpose. In the beginning his chief source of information had been the numerous articles and studies published by the Swede Bo Lundberg. Between 1960 and 1967 they had developed very serious ecological and economic objections to the use of supersonic aircraft, but they had gone unnoticed by the American public. However, the official reports of US government departments, of the research laboratories working for them and of the Academy of Sciences, if they are compared and their omissions and contradictions examined, can lead to very interesting conclusions about the attitude of the government.

Because of these reports, then, it became possible to expose the half-truths put out by the authorities and their think tanks, to penetrate their smoke-screens, and to confound, by careful calculation, their claims that the noise and boom of the SST were harmless. For instance, Shurcliff was able to calculate that if the whole of world air traffic went over to SSTs, the damage done daily would amount to $24 million worth.

Shurcliff was able to overcome the natural publicity-shyness of most scientists. His sister, Mrs Ingelfinger, the wife of a well-known doctor, regularly distributed xerox copies of the latest 'anti-reports' by scientists, economists and political experts to at least two hundred leading American newspapers, to fifty columnists and as many radio and television networks. She supplied information and arguments to numerous Congressmen, and to the authorities and commissions concerned with SST. She also sent her 'counter-information' to leading members of the Boeing Corporation, and to the management of most of the big American and European airlines that had been offered SSTs while they were still on the drawing-board.

Backed by expert evidence from biologists, physicists and doctors, Shurcliff introduced the hitherto neglected factor, 'man', into the mainly technical calculations. He showed, for instance, that a single flight from New York to Los Angeles could affect the health of no fewer than ten to forty million people. In his submission to Congress, General Maxwell of the Air Force tried to invalidate this claim by saying that his men at Edwards Air Base had lived a long time with 'sonic boom' and had got so used to it that they did not even notice it any longer.

Shurcliff demanded that the decision as to whether or not to use a new invention should be made by taking account of vulnerable civilians instead of tough soldiers. He showed me his list of vulnerable persons. It included heart sufferers, children, sufferers from insomnia and from digestive and nervous diseases, pregnant women and women in labour, and particularly anxiety-prone people. Another long list was devoted to 'special situations' where the sudden boom from the skies could be disturbing or dangerous, such as theatrical performances, concerts, difficult eye and brain operations.

Such factors are hard to measure exactly, and for that reason they had been neglected – a normal proceeding among technical planners. Shurcliff was able to show that the supposedly harmless annoyances to the ordinary citizen became intolerable when they accumulated. It is not enough to consider each industrial side-effect by itself; taken together, they constitute an overplus in terms of irritation. One needs to consider the total effect to realize that the daily harassment becomes excessive and dangerous.

Shurcliff told me: 'Sometimes pure chance came to the aid of our campaign. Gordon Bains, director of the nation's SST programme, was telling newsmen that many persons who claimed their property had been damaged by sonic booms were only imagining the damage. "I believe there's a great deal of psychology in this," he explained, when – WHAM! – a jet fighter pilot in an F-104 broke through the sound barrier at only five hundred feet. The booming shock wave which followed blew out two seven- by twelve-foot plate-glass windows. There was a moment's embarrassed silence, but then an explosion of laughter swept Bains from the platform.'

This was just an amusing episode in a dispute which grew more and more acute, and developed into an argument about the whole principle of technological progress: its direction, its finance, and its dependence on political power. Never before had there been any public discussion about major American projects such as the atom bomb, long-range missiles, the decision to use nuclear weapons and to test more powerful ones, or the decision to land on the moon – which later turned out to have been taken by not more than twelve men. The case of the SST, on the other hand, became a new model for the democratic criticism of progress.

It soon became apparent that the Nixon administration was not at all anxious to have an open inquiry. On the advice of Dr Lee DuBridge, who was the chief White House adviser on scientific affairs at the time, a commission was appointed to inquire into the critical objections to the superjet. Its findings supported the criticisms; but the White House disregarded them, and in its usual armaments-race style demanded that the superjet be built. Otherwise, it was said, America's leading position in civil aviation would be usurped by France and Britain with Concorde or even by Russia with the TU-144. Several politicians, such as Senator Proxmire and Congressmen Reuss and Yates, made repeated demands to have the commission's report published, but they were refused on the grounds of 'executive privilege'.

But now the opponents of SST were to gain a valuable ally in the person of one of the members of this secret group. It was the kind of incident that is typical of these disputes. The chairman of the White House study group, the forty-two-year-old IBM physicist Dr Richard Garwin, declared in public why, after mature consideration, he was against the SST project. Every sort of pressure from above had been put on him to keep silence, but in vain. This deserter actually became a leader in the anti-SST campaign. DuBridge turned out to be less firm. In March 1969 he too had declared himself against this 'exciting technological development' because it was 'harmful to the environment, noisy, unattractive and unpopular'. But a year later he had been brought to heel and told Representative

Yates that he was a soldier; the President had made the decision and he, DuBridge, would support it.

Meanwhile other important witnesses had appeared who proceeded to address new and even stronger anti-SST arguments to Congressional commissions as well as to the mass media. The biggest impact was made by two experts from the Council for the Quality of the Environment: they foresaw consequences which went far beyond damage to health and property. They said that the massive use of SST might, for instance, cause the polar ice-cap to melt, which would lead to floods on a Biblical scale; it might also result in widespread damage to plant and animal life due to an increase in ultra-violet radiation.

When these statements by Dr Russell Train and Professor Gordon McDonald first came over the teleprinter they sounded too sensational to be believed. They spoke of 'holes in the sky' caused by the SSTs tearing through the atmospheric skin of the earth. But when, in May 1970, the text of the statements made to Senator Proxmire's Joint Economic Committee was published, these fears began to be taken seriously.

Train had said:

> The supersonic transport will fly at an altitude of between sixty thousand and seventy thousand feet. It will place into this part of the atmosphere large quantities of water, carbon dioxide, nitrogen oxides and particulate matter. . . . A fleet of five hundred American SSTs and Concordes flying in this region of the atmosphere could, over a period of years, increase the water content by as much as fifty to one hundred per cent. . . . There is a possibility, which should be researched, that subsonic jets have been contributing to this increase.

According to Professor Howard Johnson of the University of California at Berkeley, a major increase in condensation would damage the protective layer of ozone in the outer atmosphere. This layer acts as a natural filter to protect the earth and its inhabitants from too much ultra-violet radiation, and it would lose that function. Biologically harmful wavelengths would reach the earth's surface; temperatures would rise and cause a sort of worldwide sunburn of devastating intensity among plants, animals and human beings.

Vehement discussions followed this statement, and the impression it had made was strengthened when the opinion was put forward that if the layer of ozone were to be diminished by as little as 5 per cent, eight thousand additional cases of skin cancer might be expected to occur annually among American whites.

The warnings about the consequences of SST were presented as possibilities, not certainties; nevertheless, they can be taken as the chief reason why a majority in the Senate refused to support the project. The best version of this veto comes from Professor McDonald, who formulated it with the utmost scientific caution:

> But I must emphasize we are just beginning to understand these consequences. It is a very iffy subject. . . . It would be my judgment as one who has worked in the field that the effects would be probably minor, but I would not want to take the risk of tinkering with the upper air without more information.

Two years later this cautious view was confirmed by a new inquiry conducted by the National Academy of Sciences. Their study group was even more doubtful, and recommended that SST should on no account be used for scheduled flights until more research had been done into the possible effect on ultra-violet radiation.

The difficulties of technology assessment

The battle over SST was a turning-point in the history of the scientific and technological revolution. It was the first time man had refused to produce something that he was capable of producing.

Hasan Ozbekhan, the imaginative and high-spirited son of a Turkish pasha and one of America's leading planners, formulated the new maxim that we should not do everything that we are able to do.

But how and by whom are such decisions to be taken? How can they be disinterested? Who can check that they are not abused?

These and other questions were discussed in connection with the creation of an Office of Technology Assessment as proposed by Emilio Daddario. The debate about SST had convinced not only Congress but the public as a whole that a check of this kind was necessary. But it soon became apparent that SST was a comparatively simple case whose possible negative consequences were relatively easy to spot. What about innovations whose effects may seem welcome and beneficial at first, but whose secondary and further consequences may turn out to be unexpected and problematical, if not positively harmful? What about the extension of human life by hygiene and drugs? It has already resulted in overpopulation in certain areas.

Joseph F. Coates, who works on questions of this kind for the US National Academy of Sciences, has taken two typical achievements of technology, and drawn up lists of their consequences:

	AUTOMOBILE	TELEVISION
First-Order Consequences	People have a means of travelling rapidly, easily, cheaply, privately, door-to-door.	People have a new source of entertainment and enlightenment in their homes.
Second-Order Consequences	People patronize stores at greater distances from their homes. These are generally bigger stores that have large clienteles.	People stay home more, rather than going out to local clubs and bars where they would meet their fellows.
Third-Order Consequences	Residents of a community do not meet so often and therefore do not know each other so well.	Same as left. (Also, people become less dependent on other people for entertainment.)
Fourth-Order Consequences	Strangers to each other, community members find it difficult to unite to deal with common problems. Individuals find themselves increasingly isolated and alienated from their neighbours.	Same as left.
Fifth-Order Consequences	Isolated from their neighbours, members of a family depend more on each other for satisfaction of most of their psychological needs.	Same as left.
Sixth-Order Consequences	When spouses are unable to meet heavy psychological demands that each makes on the other, frustration occurs. This may lead to divorce.	Same as left.

Coates develops his chains of consequences in terms of human relationships in a very simple fashion which could be applied to a multitude of other innovations. 'Cross impact analysis' is a more sophisticated method for tracing the interaction of different technological and social developments. But the number of possible and

probable consequences soon grows so large that it would be hopeless to try and check them all.

Some idea of the size and complexity of the task can be gained
from the duplicated questionnaires that have been produced for it:
they contain thousands of boxes, each of which represents one possible
consequence resulting from one or several procedures. One such
questionnaire with no fewer than 8,800 boxes was sent out by the
geological study section of the Department of the Interior to check
the environmental effects of new technical installations. And even
then not every criterion was represented. 'Beauty' and 'unspoiled-
ness', which mean so much to many conservationists, were not included because they were considered 'too difficult to assess'. And
yet precisely these criteria played a particularly important part in
the opposition to the Alaska Pipeline which is to connect the newly
discovered arctic oil-wells in Prudhoe Bay with Valdez, the ice-
free port on the south coast.

The assessment of pharmaceutical, medical and genetic techniques
presents new problems which are particularly difficult to solve im-
partially. For here direct personal effects are to be expected: we shall
have to question assumptions and habits that make it hard to reach
a cool assessment. What, for instance, should be the life expectancy
of the statistically average person? If it is prolonged, shall we not
find marriage intolerably protracted? Is it right that the transfer of
money and power from one generation to the next should be post-
poned by such a prolongation?

Questions like these put too great a burden not only on our
powers of imagination and sympathy, but also on our moral standards of judgment. And yet hitherto insoluble problems like these
could lead to the development of qualities such as foresight, lucidity
and social imagination which are not much encouraged in our
society. One of the reasons why these qualities have been neglected
may be that we have never needed them as much as we do in our
present situation. Mankind at the millennium will have to develop
its powers of prognosis as well as its constructive imagination.

But in spite of these difficulties, we shall have more and more
study groups to examine prophylactically the possible effects of
scientific and technological inventions. The question of their im-
partiality and independence is bound to arise. Who is to choose the
members of these tribunals for new inventions? Who will guarantee
their objectivity? How can wrong decisions be reversed? This
opens up a whole new field not only of law, but also of political
planning. Conflicts of interest such as we experience at present will

be thrashed out in advance, and this may help to make them less acute, for it is easier to agree on a future development than on something which already exists.

And future conflicts are already in preparation. Economic and national power élites are trying to 'secure the future' by using expert evidence about technological developments to obtain new concessions which will make their own position safe for a long time ahead. This is what is feared by a large part of the Third World. They are afraid that their own modernization programmes will be curtailed in the name of saving tomorrow's world. Another possibility is that industries forbidden by the dictates of technology assessment will be exported from the developed countries to more thinly populated areas.

Robert Feldmesser, a leading methodologist of technology assessment, has warned those responsible for future decisions about inventions and industrial innovations to consider the interests not only of their employers but of 'every relevant social group'. It is a high-minded warning, but it is unlikely that such well-meaning exhortations will carry much weight in practice. We are much more likely to see a tough struggle of interests developing over the introduction of each innovation. The 'victims' will not be able to rely only on the 'goodwill' of their opponents; they will need their own well-informed and influential representatives to take part in the discussions and decision-making.

Nowadays most innovations are adopted by governments or management without the public being informed. In the rare cases where technology assessment is being tried – for instance in France and Sweden – the spokesmen for the establishment are always much better informed and prepared than the workers or other unorganized groups who are therefore incapable of playing their proper part in the discussions. The old, the young, children and tenants usually have little influence, and they are the ones who are pushed aside by technological advances and have the quality of their lives impaired.

In spite of long years of resistance by industrial lobbies, the Office of Technology Assessment (OTA) of the US Congress was finally established in 1972. It keeps in touch with public interest groups, but puts special emphasis on the fact that it does not intend to become the instrument of these groups of citizens. So far it has been fairly obvious that the influence of this new authority upon technological development is to be restricted as much as possible. The sad conclusion that the mountains have once again given birth to a mouse is confirmed by the following facts: the authority's premises on the seventh floor of the old Immigration Office are very modest by Washington standards; the budget is small and by no means guaranteed for any length of time; the officials say that their studies are of a 'harmless nature'; and Congressman Mosher, the OTA's chief spokesman in Congress until his retirement in 1976, was always anxiously emphasizing that its intention was not to make proposals, still less to control what was being done, but only to put more information at the disposal of the decision-makers if they expressly asked for it. At the annual conference of the American Association for the Advancement of Science at the end of January 1974, Congressman Mosher not only described the OTA as a child that must not annoy those who feed it, but also said that it must keep a 'low profile'.

If technology assessment is not to become the instrument of an even stronger 'control from above', it must, from the very beginning, be in the hands of the public and not in the sole custody of study groups belonging to the establishment. It will be important to remember that the right of consultation should not depend on competence alone, but also on the degree to which an individual or group will be affected by whatever development is under consideration. The criterion of expertise – and in the communist countries the criterion of political indoctrination as well – can too easily be used to give the expert or ideologue an undemocratic advantage.

Expertise is admirable when it is used to inform the ordinary citizen, but not if it leads to superiority, conceit and fresh privileges. At the turn of the millennium the unqualified will hardly continue to tolerate the privileged position of the expertocrats. The physicist Max von der Laue has already warned scientists that if they neglect to establish contact with the people they will one day end up hanging from the lamp-posts.

There is one group that should be represented by a statutory advocate before the tribunal on new developments: the unborn. They are the ones who will be affected by the long-term consequences of technological innovations. Many of the things that researchers find and engineers build will change and endanger not only space, but time as well. If such advocates had existed in the first half of the twentieth century, they might have been able to protect their clients from the shameless waste of their irreplaceable heritage, from the squandering of their rich capital of raw materials and other natural wealth by exploiters who wrongly considered these things to be their livelihood, and who were blind to the future. But today the protection of the future and its inhabitants is even more urgent than it was fifty years ago. For now we are concerned with their biological heritage, threatened by the increase of dangerous radiation. The earth may even cease to be habitable because of the irresponsible advance of technology. The defenders of the unborn must not cease to insist that we hand down as much as we can of the wealth, the beauty and the opportunities for a good life on this planet.

Soft technology and critical science

This is the point of departure for those who think that it is not enough to tame the old technology, since that would probably be impossible. They want to create a new 'soft' technology which will be less hostile to life.

One of the chief spokesmen of this trend is an intelligent and courageous English writer, Robin Clarke. When I first met him he was chief editor of the much-respected London monthly *Science Journal*. That must have been towards the end of the sixties. In those days he still believed that, because of their rising social status and their objective thinking, scientists would be able to lead the world out of its great crisis.

The next time I saw him, he was no longer in his editorial office but in the large, light UNESCO building in Paris. He had a small brown beard and seemed more relaxed and open. Round his desk four young men dressed in the fashionable revolutionary style of protest were talking about a coming conference of critical-minded young scientists to discuss the environment.

Clarke's journal, so he told me, one of the dozens published by IPC in London, had been closed down from one week to the next, not because it was unsuccessful but because the owners considered it insufficiently profitable. Presumably this had helped to change Clarke from an enthusiast for progress into a sceptic. The material for the book he was writing on the arms race had also helped to sharpen his critical eye. He no longer thought the scientists very influential or very sensible: he found them over-specialized and much too dependent to fulfil the demand for truthfulness inherent in their calling.

Of course there were exceptions, and some of these had chosen Clarke as their focus now that he was a leading official in the Science Division of UNESCO. The scientific and technological underground that he gathered about him had developed, intensified and politicized the ideas of the disarmament campaign of the preceding decade, whose leading members had been atomic physicists. The older generation had hoped personally to influence the statesmen of East and West; these younger men were more sceptical.

Their criticism went deeper. For them the root of the trouble lay in the economic power structure, in the democratic institutions which had grown pusillanimous and therefore stagnant, but most of all in science itself and the way it was alienated from life in the service of an ever more dangerous technology. One of these new-type scientists was the nuclear physicist Dr Philip Noyes, employed on research in the acceleration laboratory at Stanford University; he publicly accused the President of the US of committing war crimes by means of research. Another atomic scientist, Dr Charles Schwartz, proposed to his students at the University of California that before attending his course they should swear a solemn oath never to use what they were about to learn for inhuman purposes.

Various groups of critical research scientists and technologists were formed during the sixties, especially in Holland, France, Sweden, England and the US. Only some of them were Marxist in tendency. Their deliberations pursued a new course. They felt they could see a ruthless, dominating, anti-life attitude in the very way in which research and development treated the world as an object to be analysed, X-rayed, dissected and manipulated. The opponents of tyranny and oppression should ask themselves, they said, to what extent scientists, by their very style of thinking and working, were responsible for the destructive tendencies of the age.

Peter Harper was typical of this new attitude. He had been a neurophysiologist at Sussex University, but abandoned overnight a promising career in brain research because he thought his work might easily be misused. Shortly after this he helped to found the New Science Group, which met for the first time in March 1971 in the Mental Health Research Institute at Ann Arbor, Michigan. Its aim was to formulate 'a new moral philosophy for scientists' which was to challenge the hitherto almost unquestioned assumption that science was objective, rational and neutral. This self-criticism was to be the first step in formulating new concepts and goals for research and development; these were to be more modest, more open and more committed to mankind. In a provisional nine-point programme the group called for a blueprint for new Utopias based on new science. As their most urgent goal, for immediate practical action, they cited 'soft technology – hopefully without risks. Use of natural materials which regenerate themselves; minimum use of energy.'

A year later I visited the Stockholm conference on the environment. By this time soft technology had gained so much ground
that it was able to mount its own exhibition, using the old American
Indian word 'pow-wow' for its title. It occupied a long pavilion
on the island of Skepsholmen which was filled with descriptions
and models of experiments for making use of the sun, the wind,
river currents and ocean waves. There were photos, sketches and
dissertations dealing with 'unaggressive' methods of agriculture
and new types of tools and machines less hostile to the environment;
and there were reports from research stations, mostly student
communes in North America and Europe.

It was all very sympathetic, but the bays with their mostly rather
primitive or improvised-looking plans and blueprints, their hand-
written wall newspapers, their little piles of pamphlets, prospectuses
and manifestos produced an impression that was bound to repel
rather than attract the exhibition visitor used to a high degree of
typographic perfection. It was impossible to shake off the idea that
these were enthusiastic but not quite serious amateurs. Pitted against
an industry that knows how to project a serious, detailed and
powerful picture of itself, they probably would not stand even an
outside chance. Or did they really believe the claims made by their
fellow-campaigner Harold Bate? He had achieved a brief moment
of fame through a British television broadcast, and now he maintained that the motor industry had been thrown into panic by his
methane gas engine which was powered by heated chicken-drop-
pings. He told visitors to the exhibition that motoring journals had
refused to print advertisements for his methane converter. It would
have meant losing thousands of pounds' worth of advertising from
the oil companies. He had been told that if he lived in America he
would have been quietly wiped out long ago.

I had expected more, too much, perhaps. The theoretical possibilities of soft technology went far beyond these well-meaning do-
it-yourself exhibits. Robin Clarke must have sensed it. He did not
even come to Stockholm to see this pathetic spectacle. Three months
previously, we had been on the same plane to Washington and he
had told me that serious soft technology still needed a great deal of
patient development. 'Perhaps the technology that I am sponsoring
will never work in practice,' he ruminated. 'It was all right for the
primitives. But we must aim at something much more delicately
organized, much more sensitive, and with many more nuances.'

But surely that will require a quite different sort of person from the
eccentrics who follow this as they do every other movement for

innovation and reform? And how can a new technology financed only by the limited means available to a few private enthusiasts hope to present a serious alternative to hard technology, supported as it is by existing legislation, existing habits, enormous sums of money and innumerable specialists?

In my slightly depressed mood I ran into Philippe Boitel, a twenty-year-old Parisian with the usual Jesus hairstyle of the underground, flowing white robes, and an expression that kept changing from gentleness to burning enthusiasm. However, he turned out to be more sceptical and intelligent than his appearence might have suggested.

He was kneeling on the worn grass in front of the exhibition hall trying to mend a large plastic bag. Every so often he would get up to pump air into the transparent skin, which would swell and then collapse like a huge sleepy animal.

'*Merde*,' he swore. 'Another hole. But we'll fix you all right.'

I asked what he was doing.

'Wait a moment, wait a moment. There – now!'

He seemed pleased with his last effort, and dragged the formless object to the water's edge.

A handsome long-haired young man in canary yellow trunks was standing impatiently on the landing-stage. 'Have you fixed it?' he asked sullenly. 'Can I try it again?'

'Yes, I think it's OK now. There, it's holding air.'

Meanwhile the plastic bag had reached its full and impressive size. It floated on the water and only rocked very slightly when the sulky Adonis got into it as though it were an ordinary boat. And then he began to move forwards on his thin cocoon. Yes, he was walking on the water. He remained upright, gliding and floating across the bay. All the spectators standing on the shore burst into applause as the boy clad in a cloak of spray danced into the afternoon sunshine.

But then a wave from a powerboat hit him. He swayed, regained his balance, flexed his knees three or four times. The fifth time he keeled over.

'But it was fun, wasn't it?'

'That's going to mean a lot more work,' sighed Philippe.

'Well, OK. It's a lovely toy. But I thought you wanted to save the world.'

'Man, you can only do that if you have fun. You can't do it with your teeth clenched.'

That was how I got my first lesson in soft technology. It is made not only with different concepts, different intentions and different materials, but also in a different mood: without haste, without fear of failure, without grim ambition: with gaiety and joy.

Then is it just a hobby after all? Just an amusing pastime?

'Up to a point. No. It's more than that,' said Philippe.

He had just returned from Abyssinia where he and a young American, John Morgan, and their team had lived with the Ethiopian peasants. There these inventive and selfless development-helpers had started a 'village technology innovation experiment': its object was to bring knowledge and help to villagers all over the world without upsetting their traditional way of life.

Philippe recounted: 'We learned at least as much from the villagers as they from us. We gave them more and better varieties of seed, a few simple, easy tools and a few fairly useful ideas. For instance, how to collect water in the rainy season. Or what to do if you want to make methane gas as well as manure from waste products. Mr Bate's invention didn't seem at all silly to us then. And they taught us that work can be a pleasure if it's a challenge: your own strength, your own imagination, your own brains. And that's where small-scale technology fits in so much better than the vast, 'perfected' methods, which are fun only for the inventors and perhaps the makers of the first prototypes: the people who have to mass-produce them sweat away and get bored.'

Small is beautiful

Ideas like this have led a few planners in the Third World to design their own models, no longer under the inspiration of the great industrial powers. M. S. Iyengar, the head of the North Indian Research Institute at Jorhat (Assam), sees it as his task to find new kinds of technological possibilities for a subcontinent already suffering from overpopulation and famine.

His work led him to conclude that industrialization in the usual Western style would be dangerous rather than beneficial for India in the long run. The huge new steel works and modern factories built with foreign help act like magnets and lure thousands of people from their villages. But what is to happen to them, asks Iyengar, when production has to be rationalized and automated in order to compete on the world market? In the developed countries they could be absorbed into the service industries, but here they remain unemployed and turn into the lumpenproletariat that you can see vegetating in Bombay and Calcutta.

'We need a quite different technology from the one you are

offering us,' Iyengar declared at the second world futurology conference in Kyoto. 'Not giant combines but lots of small power
stations and workshops scattered over the whole country. Industry
must come to the villagers, not they to it. And industry must not
try to save on labour. On the contrary, it must produce more and
more jobs. Every Indian has a right to find a productive occupation.
Therefore our plans for the future must concentrate less on getting
better performances from our machines than on giving satisfying
work opportunities to as many people as possible.'

As a result of this changed attitude, Indian interest in large-scale
industrial development models is diminishing. Maoist planning
models are gaining in influence, though not so much for political
reasons. Influential planners feel they can use the decentralized
methods of Chinese industry without bothering about their ideological background.

Similar ideas are gaining ground in other countries of the Third
World. The more recently independent ones especially agree with
the well-known French development specialist René Dumont.
After many years of experience he wrote his book *L'Utopie ou la
mort* (Utopia or death: Paris, 1973), in which he concludes: 'Yes,
the Chinese experiment seems the society most likely to offer an
assured future.' It is based on the Cultural Revolution principle of
'independent industrial systems' (*kung-yeh-ti-hsi*). Numerous small
and medium-sized local factories were started after 1959, and more
particularly after 1966. They are proud to model themselves on the
much admired workers of Tachai, to stand on their own feet, and
to be wholly or almost wholly independent of outside help. A
typical sign of this trend is the absolute and relative increase in the
numbers of very small local power stations built by the people
themselves. There were nine thousand in 1960 and thirty-five
thousand by 1972. At the Stockholm conference on the environment,
the Chinese delegation told Western environmental activists that their
country's policy was deliberately directed against large-scale industrial
plants in order to prevent uncontrolled growth which would make
it impossible to provide work opportunities for each individual.

The British Cabinet Minister Anthony Wedgwood Benn has
described the relationship of the young Chinese to their technology
as follows:

Here is a society trying to make scientific and academic
knowledge work more directly for it, and debates about that
are raging world-wide.

Perhaps the most vivid explanation of what all this means came to me when I visited the factory in a rural commune near Peking. There, in a barnyard workshop, I was shown a machine tool that had been assembled by four teenage boys who had, until a few months before, been making harnesses for the farm horses. Recalling many ministerial visits, I expressed an interest in the machine I was being shown. But it was not the machine that interested them.

What they wanted to tell me was that in the process of assembling the machine these boys had developed an ability and revealed a talent and demonstrated a self reliance that would never have been possible before the Cultural Revolution. So I found myself smiling at the boys and they were smiling at me and both of us were thinking of the machine only as an instrument of their development and as a tool they could use but not as having any intrinsic merit in itself. It revealed an attitude to technology that we are learning rather late.

Even in the West people are beginning to rethink along these lines. One of the most ingenious and influential minds behind this trend is E. F. Schumacher. He emigrated from Germany to Great Britain in 1930 and eventually reached the boardroom of the nationalized coal-mining industry. In 1965 he began to concentrate on a project of his own that had grown out of experiences and insights he had gained in Asia. It is called the Intermediate Technology Development Group.

This group is opposed to the idea of trying to help the Third World with large-scale technological projects which destroy its way of life and make it dependent on the industrially developed countries. Instead, they believe in exporting to underdeveloped countries only tools of a kind that will give the inhabitants occupations of their own and opportunities to develop initiative: such things as pumps, welding equipment, small motors and generators.

Schumacher gave the following example of his unorthodox attitude towards the problem of helping undeveloped countries. A newly independent African state was trying to obtain a modern rotary press in England in order to produce massive editions of the daily paper published in its capital. If this machine had been supplied, the whole country would have got its daily news from one source, and the way would have been paved for complete uniformity; at the same time the cultural variety among the West African tribes would have been endangered.

The Intermediate Technology Development Group therefore proposed that several small and medium-sized printing presses for small and medium editions should be sent out instead. This improved the information services without imposing total uniformity.

A factory which opened in Dakar a few years ago for the mass production of plastic sandals may serve as a warning against mindless industrialization. Its output was sufficient to supply the whole country. Hundreds of native craftsmen who had made shoes from bast, hemp and leather were ruined, and the Senegalese were forced to replace shoes that were both beautiful and suitable to the climate with cheap rubbish.

'Small is beautiful' is the formula Schumacher invented in contradiction to the American maxim about the beauties of size. It has a future in an age that is beginning to turn away from gigantism and enforced uniformity. In the developing countries especially a definite change of values is becoming apparent between one generation and the next. People between the ages of forty and eighty still regard the showy megamachine as the height of prestige and the goal of ambition, while younger men have long since rejected it.

I found this change of attitude confirmed by an international gathering of Christian youth in the Finnish town of Turku which was chiefly attended by members of the developing countries. Brazilian, West Indian and Indonesian students agreed on the supreme importance of developing their 'own' technologies suited to the economic, social and human needs of their compatriots.

'Our fathers still think differently,' they said. 'But we know that it is nonsense for us to import sophisticated labour-saving machinery when we have an excess of unskilled labour. It costs a great deal of valuable foreign currency and makes us dependent on imports. Our people have a different work rhythm because our climate and our temperament are different. We have had enough of the whites forcing their pace on us. . . . Most Western machines quickly deteriorate in our countries. We ought to build machines from our own materials. . . . Western technology has given us monoculture and giant slums. It claims to make you rich, but in fact it only makes deserts and human ruins.'

Professor Henryk Skolimowski, who considers himself one of the 'new' sociologists, has heard similar comments. 'You whites are crazy,' an American Indian said to him. 'Your machines drive you to work like slaves until you are sixty. Then, when you are already too old for it, you begin to fish, to travel, and to enjoy yourselves.

We spread the hours of pleasure evenly over our lives. Isn't that more sensible?'

Visitors to North Vietnam have been able to observe a new indigenous non-Western technology which uses local materials and is based on local handicrafts. The natives of Swaziland had different problems to cope with: they solved them by developing their own pumping machinery suited to local conditions. Workers in the Euphrates delta refused the prefabs they were offered and invented a new process of building with reeds. Japan used to be the classical example of the imitation of Western models: now it is beginning to develop a specific 'Nippon technology' based on electronics combined with the traditional high-quality handicrafts.

The development of specifically Asian, African and Latin American technologies is only just beginning. What they have in common, however far apart they may be geographically, is the desire to be closer to life and to nature. The reason is not hard to find. They spring from the protest against the mechanistic, insensitive, standardized technology of the West with its emphasis on speed and maximum output. It is not unthinkable that before the turn of the millenium the West will be sending for black, brown and yellow technologists to come up to the citadels of industry and show their former teachers how to produce the necessities of life without waste, without harm to man and his environment, without haste and without alienation.

Experiments with soft technology

But let us return to the efforts to make soft technology not merely a romantic flight into the past but a more humane path into the future. Peter Harper has described various strategies for developing soft technology. Only one of them has been tried so far. It consists in testing existing or historic technologies to see if they are worth employing or, in some cases, re-establishing. There are building materials, for instance, that have been dropped because they were thought too primitive: clay, earth, stone and wood. Further to this, other materials and processes that could be described as 'soft', or biologically sound, are to be brought into future experiments: building materials based on the soya bean, new forms of ceramics and glass. Increased efforts are being made to use 'old' sources of energy such as the sun, water, wind and hot springs; successful experiments with the latter have been made at Laredo in Italy. When it comes to food supply, we should pay attention not only to the return to the use of organic manure, but also to experiments in 'hydroponics' (growing plants without soil in water impregnated with chemicals) and to the extraction of protein from leaves and plants hitherto considered inedible. In transport, we should among other things try fresh experiments with noiseless aircraft such as balloons and dirigibles.

A second strategy would define 'essential inventions', find new solutions, and develop them where they already exist in embryo. Under this heading come the development of solar cells and solar power plants; the conserving of solar energy for sunless periods; 'soft' printing methods like xerography; better processes for making high-grade steel in small quantities; economic methods for recycling waste; the improved durability of natural fibres; new cooling techniques; and so on.

This list does not include an idea that is continually being promoted by the Englishman Oliver Wells and others. They maintain that it would be possible to develop techniques for making various products in germinal form: these 'seeds' could then be raised like plants by a process similar to photosynthesis. Presumably Harper and his friends keep quiet about these and other wild-sounding ideas because they do not want their efforts to be written off as science fiction from the very beginning. The second group of their ideas in particular shows that they are chiefly concerned with. finding practical transitions from the present large-scale technology to a self-supporting kind of soft technology.

Nevertheless, we also hear of a third and more daring strategy to develop and test far-reaching 'social Utopias'. One of the central questions here is how to convert millions of people, and not just a few communes of 'extremists', to soft technology. The central problem is this: how can modern man, accustomed as he is to the comforts of mass production and mass consumption, be made to accept the more modest and physically more strenuous lifestyle of soft technology, and what kind of satisfaction could he derive from it? These things cannot be decided by writing books or holding discussions. People must be ready to sacrifice their bourgeois existence in order to test whether such a life-style is even tolerable. The possibilities and problems of soft technology indubitably need to be tested through social experiments.

Robin Clarke, who has already been mentioned, gave up his highly paid tax-free job as an international civil servant in Paris because he felt that in spite of his high salary he had hardly any time left for himself and his family. Life in the French capital was expensive and exhausting; it gobbled up his income and ruined his health and *joie de vivre*. In addition, he suddenly felt sick of writing and speaking about 'new science' and 'new technology': he wanted to try alternative living in practice. So he banded together with a few like-minded spirits, most of them from professions 'engaged in manipulating with symbols' – publishing, television, the universities. They founded BRAD (Biotechnic Research and Development), and settled as a commune in a part of Wales where land was not too expensive.

BRAD is trying to prove that even comfort-loving modern man can live without robbing or destroying the environment. The community gets its energy from a nearby stream with the help of a heat pump. Later on they intend to build simple solar collectors and a windmill. They are going to try to develop storage methods for sources of energy that are partly dependent on the weather. That is one of the first research tasks they have set themselves. On the model of the New Alchemy Institute at Cape Cod, Massachusetts, they are trying to establish an autonomous nutritional cycle: organic waste is fed to fish and cattle and so brought back into the natural cycle. Clarke has worked out that forty-two acres of land farmed in this way would comfortably support forty people. If roughly one acre is sufficient to feed one person, then millions of Britons could live like this without hardship.

The first months at the 'green laboratory' showed that comfort-loving intellectuals can adapt to this kind of life. Only experience will show whether it can continue to work. It has already become evident that the experimenters have underestimated the difficulties of group dynamics. *Undercurrents*, a London periodical with the subtitle 'The Magazine of Radical Science and People's Technology', reports as follows:

BRAD . . . prefers to be known these days as 'Eithin-y-Gaer', the name of the Welsh farm where the group has settled. The change of the name symbolizes the change of emphasis which the community has undergone over the past couple of years, a change which culminated in Robin and Janine Clarke's withdrawal from the community. Robin apparently felt that the group's role should remain largely as he originally envisaged it at the beginning . . . the others, however, felt that they needed to get their heads and their own inter-personal relationships straightened out a good deal more before they would be ready to start telling the world what to do.

It is probably sensible that the members of BRAD return to their occupations and life-styles in the cities for three months out of every twelve. There are two additional disturbing elements they have to cope with. The first was unexpected: many things that the neo-technological community would like to make are forbidden by the authorities. The other is a question which preoccupied Clarke even before he took the leap out of his bourgeois career: is it right for people who feel responsible towards society to retire from it and 'cultivate their gardens'?

But he does not regard his experiment as a flight from the world. One of the chief motives of this former international administrator was the hope of having an effect on the Third World. He thinks that the new slogan of the industrial countries, 'Put an end to growth!', will be more readily acceptable if at least some of their inhabitants practise what they preach to the rest of the world. He hopes for a new 'convergence': the condition of the 'poor' countries will be improved by a new, sensible, soft technology geared to their own needs, while at the same time the 'overdeveloped' countries will gradually start reducing their luxury standards of consumption.

Peter Harper mentions yet another motive of a social kind for experimenting with soft technology as soon as possible. In common with other technologists such as Professor Donald N. Michael of the University of Michigan, he is alarmed at the continual increase in the size of things. Many products of large-scale technology such as aircraft, tanker ships, power stations, etc., are continually becoming more bulky and more complicated. This increases the difficulty of maintaining them. The possibility that one of a thousand parts – or, in the case of really large systems, one of a hundred thousand parts – will go wrong grows daily greater and more dangerous. Peter Harper therefore demands a new special science which will deal with the causes and effects of man-made disasters. Among the cases he has already studied are the following:

◇ the mercury poisoning of Minimata Bay in Japan;
◇ deformities caused by thalidomide as well as other diseases and deaths caused by insufficiently tested drugs;
◇ the break-up of the tanker *Torrey Canyon*;
◇ the killing of fish in big rivers (Rhine, Mississippi, etc.) through oil leaks;
◇ the atomic reactor accidents of Windscale, Detroit, Boulder;
◇ the spread of epidemics (schistosomiasis) through artificial reservoirs (Aswan dam, Kariba dam);

◇ the 'greenhouse effect' created by changes in the weather caused by an increase of coal gas.

Harper admits that the safety regulations for the prevention of technical disasters are improving all the time, but on the other hand the number and size of potentially dangerous constructions is on the increase. Even if we can count on a statistically smaller accident rate, the scale and effects of each single accident are presumably growing larger all the time.

This theory can be demonstrated even today if we look at the development of civil aviation. It is true, as the airline companies claim, that the accident-free mileage flown increases almost every year. But in spite of this the annual number of deaths through air accidents rises. How can this be explained? Well, there are more and more aeroplanes in the air, and each year the size of passenger planes increases. Aircraft carrying five to six hundred passengers are being built, and even bigger super jumbo jets for eight hundred persons and more are at the planning stage. In the future, an accident rate of 'only' five to ten passenger planes per year will cause more deaths than there are now, when more planes crash but these planes are smaller.

The dangers of 'high' technology demand new concepts and experiments to test its viability. Social and technological concepts are more closely interdependent than used to be assumed. We have particularly neglected to consider how much social conditions depend on the organizational structures imposed by technology.

Signs of degeneration in the socialist states that have been pointed out by left-wing critics are presumably to a large extent the side-effects of the centralized, repressive 'old technology' that values the product more than the producer. True, the revolution changed the ownership of the means of production, but not the spirit of production which had evolved in the days of rising capitalism: it is still directed towards output and profit, towards exploiting and adapting the workers according to its needs. The 'Trojan horse' of technology – especially of the armaments industry – is largely to blame for the failure of a humane kind of socialism to materialize. So far the Eastern bloc has not dared to design a 'technology for the people' based on the ideals of fraternity and equality. Perhaps the Chinese experiments point in a new direction.

The soft technologists keep repeating that they do not want to confine themselves to changing man's implements: they aspire to change society, but there is no chance of that happening without a new type of technology. Robin Clarke has worked out a list in which he opposes a 'community' using soft technology to a 'society' using hard technology:

Society with Hard Technology	Society with Soft Technology
1 ecologically dangerous	ecologically adapted
2 high energy consumption	low energy consumption
3 heavy pollution	light or no pollution
4 one way use of materials and energy	recycling of materials and energy
5 narrow timescale	wide timescale
6 mass production	emphasis on artisanship
7 high specialisation	little specialisation
8 'nuclear' families	extended families
9 predominantly urban	predominantly rural, or small town, communities
10 estrangement from nature	integration with nature
11 authoritarian politics	democratic politics
12 technological boundaries are of economic nature	technological boundaries are natural
13 world trade	local trade
14 destruction of local culture	preservation of local culture
15 abuse of technological opportunities	laws against abuse of technology
16 destruction of other life forms	partial dependence on the presence of other life forms
17 innovation motivated by profit and war	innovation motivated by needs
18 growth economy	zero growth
19 capital intensive	labour intensive
20 creates generation gap	brings young and old together
21 centralised	decentralised
22 productivity increases with size	advantages of small-scale production
23 processes too complicated	processes generally comprehensible
24 technological accidents frequent and serious	technological accidents infrequent and insignificant
25 totalitarian solutions of technological and social problems	diverse solutions of technological and social problems
26 monoculture in agriculture	diversity of agriculture
27 quantity receives priority	quality receives priority
28 nutrition by specialised industry	food industry involves everyone
29 income as incentive for work	satisfaction as incentive for work
30 complete interdependence of all productive units	self-sufficient small units
31 science and technology estranged from culture	science and technology form a part of culture
32 science and technology of specialist elites	science and technology practised by all

33	contrast between work and leisure	little or no difference between work and leisure
34	high unemployment	concept of work non-existent
35	technological goals for a part of the planet for a limited time	technological aims valid for all at all times

A glance at this list may lead one to conclude that soft technology is one of those para-primitive solutions recommended by Clarke's compatriot Gordon Rattray Taylor as the road to a 'new happiness'. To begin with, at least, it is a 'way back', and one wonders how many will wish or be able to take it. Critics like Michael Kenward of the English weekly *New Scientist* think that only 'a handful of intellectuals' will be able to afford such experiments, and even they will not be able to exist without high technology which will have to supply many of their tools, drugs and other aids.

But Clarke and other soft technologists see their back-to-nature movement only as a first step. In contrast to many other rural communes, they have no intention of escaping from the problems of the age and living a purely self-fulfilling life as basket-makers or weavers. They want their experiment to be a model and to test new attitudes which Clarke groups in three categories:

◇   men before machines;
◇   citizens before states;
◇   practice before theory.

He hopes that 'green laboratories' will now have learnt from the mistakes (as he sees them) that were originally made at BRAD. Science there will not be the prerogative of a specialized élite which has become estranged from real people with real needs. A more humane technology will be developed by the people themselves who need it, and for purposes which they themselves have chosen and which do not exceed their mental or physical capacities.

The pioneers of soft technology deserve much credit for submitting their ideas and proposals to constant self-criticism. In 1974 Peter Harper in particular published a number of articles in *Undercurrents* which showed that in existing social conditions alternative technologies could really function only 'under special circumstances, which restrict them to minorities with eccentric tastes, lots of money or a willingness to muck about with their life-styles. Alternative technologies are therefore either financially or culturally unacceptable to the vast majority of people and can be criticized as irrelevant, élitist or even exploitative.'

The main problem in designing soft technology for a population of billions at the turn of the century is its relationship with high technology, whose products will still be needed – at least according to present planning ideas. What would a mixed technology look like, in which huge steel and electricity works would continue to exist side by side with an important sector of industry functioning according to the ideas of the soft technologists? Harper sees it somewhat like this:

Large- and small-scale sun, wind, water, geothermal, etc. energy usage

Collective digestion of all organic wastes for methane

Total energy systems and district heating

Heat pumps

Hydrogen and methanol as basic fuels

Central electricity generation from many different sources with local grids

Fuel cells

Controlled exploitation of abundant resources

Advanced autonomous houses outside the cities, mostly of large size

All the main industrial materials (although not in overwhelming abundance)

Wide range of special steels and plastics

Semi-conductors and electronic bric-à-brac

Some computers, radio, TV

Much automation in tedious production jobs

Lithopresses

'Proper' research and development

Knowledge firmly based on the traditional sciences

Tractors, combines, machines for capital-intensive organic farming

More even distribution of population

Many more in part-time food production

Trains, public transport, dirigibles, few planes

Travel but less commuting

Strong control on environmentally harmful substances

Very strict emission standards

Less mining with improved work conditions

Controlled distribution of raw materials – careful conservation programme

Equipment designed for reliability, ease of repair, long life and recycling of components

Smaller range of consumer goods – wider range of living patterns and possibility for greater variety within a person's life . . . etc.

Back to nature, modern style

As early as 1962 the English cyberneticist Gordon Pask, writing in the New York journal *Electronics*, pointed out how strange it was that the order we impose on our environment is not an image of our own physical organization. The result has been the creation of a deep gulf between ourselves and our environment. Until recently it was still possible to live with this breach, but today we can no longer shut our eyes to the truth that – at least ideally – the second nature which we ourselves create must be an extension of mankind. The machines that we really need should be built on the basis of biological designs.

Pask's views fascinated me. Here at last seemed to be an answer to the question of how we could change technology so as to bridge the alienation gap between man and machine. I wrote to him, and received a friendly invitation to visit him at Richmond in Surrey.

The name of his laboratory, Systems Research, led me to expect a large modern institute. Instead I found a narrow-fronted old terrace house. The boss was working in the attic with a stuffed crocodile on guard. He was like something from a short story by E. T. A. Hoffman: a small, fragile, fidgety body supporting a head that seemed much too large, with shaggy hair and exceptionally pale, penetrating sorcerer's eyes.

He spoke so fast that it was hard to follow: he would interrupt himself, ask if I had understood, not wait for an answer, then begin again slightly more slowly but soon fall back into his usual pace. Suddenly he stopped, refreshed himself with some kind of brew from a thermos flask, knocked the ash from his pipe into a huge copper pot (I afterwards discovered that he allowed it to be emptied only once a year), and then asked anxiously: 'You did understand, didn't you?'

He cares about 'understanding', not just in my case but concerning his work in general. It seems to him the most important condition for its further development. He earns his daily bread by the invention and development of teaching machines which 'understand' their pupils and adapt to them. It seems possible that with their aid learning times may be reduced by a third, and this prospect appears so attractive to various industries, as well as to the American Air Force, that they have been supporting the work of this outsider.

Pask's interests extend far beyond the learning situation and its improvement. Starting from the ideas of the American cybernetic genius Warren McCulloch, he is continually experimenting with the relationships between the individual and the outside world and between individuals among themselves. What is the best way of making contact? Certainly not a situation in which only one side can be active while the other remains passive, where one commands and the other obeys – as is frequently the case between teacher and learner. 'Conversation' in which each partner receives something from the other is far more interesting and varied.

If men could talk to machines, then their hitherto disturbed relationship would change. That is the fundamental idea behind 'evolutionary technology', which grades today's machines as dull, rigid and unintelligent. Pask dreams of a not too distant future when there will be flexible, highly differentiated machines, each with its own characteristics. He foresees development in two main directions, with the 'higher kind' remaining dependent on man because of his creative faculty. A man–machine system of this sort will be based on mutual understanding and co-operation, with each side learning from the other and thereby developing further. The beginnings of such a symbiosis can be seen in the relationship between computers and their operators. It will be extended, according to Pask and others of the same mind, to as yet uninvented 'centaur-like combinations' in which a 'P individual' with human Personality attributes is linked to an 'M individual' with Mechanical attributes capable of development.

The second trend is towards 'intelligent machines' which will work to a large extent independently. Thanks to their 'electronic senses' and their ability to make decisions on information received, these technical systems will largely be able to programme themselves, so that they will be capable of functioning with little or no help from their human partners. It is thought that these machines will be receptive to rational or even to ethical concepts.

Pask compares the functioning of these networks of cybernetic machines sensitive to their surroundings with the processes inside the human body. The body too is full of constant gear-changes and productive functions which the individual does not consciously notice. We notice them only when an organ goes wrong. The day will come, Pask thinks, when the processes of technical production will be so well adapted to its 'body', i.e. the environment, that their function will be almost taken for granted and will need no special attention. In relation to this 'body' the genus man would then be the 'head'. The head thinks, invents, clarifies plans, and interferes only when there are problems to be solved.

People working on the humanization and transformation of technology are always using words like 'flexible' or 'sensitive'. In this way they reflect the attitude of Norbert Wiener, the founder of cybernetics: he saw rigid orders and attitudes as the chief obstacle to man's continued existence and development.

Warren Brodey, a Boston neurologist, was so fascinated by the work of the cyberneticists in neighbouring Cambridge that he joined their group. He sees the plant and animal world as a model for 'alternative technology'. Similar ideas are being developed by Richard R. Landers, the manager of the NASA-linked 'reliability department' in the electronics firm of Thompson-Ramo-Wool-dridge at Redondo Beach, California. He has worked for many years on artificial cells which he calls 'dyblocks'. They draw their energy from electromagnetic fields, from sunlight or moisture; they multiply of their own accord, die, and are replaced by the system that supports them.

All this sounds fantastic, but the champions of the biological development trend in technology can point out a few successful examples. Petrol tanks and car tyres that 'heal' themselves are in everyday use. There are camera lenses based on frogs' eyes; mechanical excavators adapted from the nervous system of the human arm; sonic equipment copied from dolphins. Perhaps such links between the living and the mechanical worlds foreshadow a new relationship between man and nature. The oppressor–oppressed situation, in which both sides change places from time to time, could be superseded by an alliance.

Warren Brodey calls the dream children of this kind of co-operation 'bioptems', i.e. biologically optimal structures. He sets them up as models for a new world: our present condition, he thinks, is one of clockwork mechanisms on the one hand and humans adapted to them on the other. He calls this unholy alliance a 'mechymax'.

At the beginning of his career as a practising cyberneticist, Brodey himself was a cog in a mechymax. He was commissioned by NASA to work on equipment for observing and controlling the movements of astronauts at a distance of millions of miles. It was able to record and transmit what went on in their bodies more accurately than their own senses were able to tell them. He hoped that one day this equipment would be used in a dialogue between men and machines. As in the model of its co-operation with the astronauts, its superior sensory perception could provide corrective aid for 'human enhancement'.

But he soon realized that the humanization of the machines did not mean that they were going to be used for humane purposes. On the contrary, almost all the early products of evolutionary technology, the machines with almost human and sometimes super-human capacities, are in the service of violence and destruction, from the computers in various stages of development to the 'people sniffers' for guerrilla warfare and the 'electronic battlefield' con-trolled by various robots. What else was to be expected? All this equipment has been built with money from the US Defense Depart-ment, from NASA and from the Atomic Energy Commission. Nowadays only the armaments economy can supply sufficient funds for 'exotic' technological development: the army has plenty of capital for gambling and does not need to worry about a guaranteed return.

Brodey tried to ignore this simple economic fact. He left NASA, and first with the help of MIT and then with financial help from a friend and collaborator he hoped to develop a peaceful and peace-promoting type of technology. His friends warned him it would never succeed; but he persevered.

Warren Brodey's Ecological Toys and Tools Laboratory is hard to find. It lies a long way off the main road in Nashua, New Hamp-shire, in a disused stone quarry. 'Johnson's Quarry' is the name of this area, which has been much ravaged by man. Avery Johnson, the lean thirty-year-old owner of the site, came to meet me in his station-wagon at the bus stop in Nashua. During the trip he told me that for five years he had collaborated closely with Warren McCul-loch's neurophysical group at MIT. This was a subsidiary of the famous RLE (Research Laboratory for Electronics) where biology and electronics, natural and artificial information systems, and what Norbert Wiener called 'men and men–machines' were linked in mutually fruitful collaboration.

Johnson had first met his present partner Brodey in the 'F & T Delicatessen', where the 'grand old man' Wiener had his regular table, always a centre of brilliant improvised conversation. Their collaboration developed later when they were both experimenting with new orientation aids for the blind. Blind people pick up signals from the outside world which others cannot catch; they use sounds, smells and touch to orient themselves. These facts interested Brodey and Johnson. Would it not be possible to activate these senses in the sighted too? How would increased perception of this kind affect man's relations with his environment? And would it be possible to go further and develop other new faculties and attitudes?

'Don't expect to see finished machines or equipment in practical use,' Johnson warned me as we bumped over a sodden unpaved road through thick woods. 'At the moment we are just building toys to help us think further. They are embryo forms of the kind of sensitive technology we have in mind. That thing over there, for instance.'

The trees had just given way to the shores of an emerald green stretch of water surrounded by steep, almost mirror-smooth grey walls of rock. In the midst of this was a lonely swimmer. I looked in vain for prototypes of a bioptem. All I saw was a shabby modern beach hut with windows overlooking the artificial lake. In front of it was a sort of swing, a large board hanging from four ropes and gently swaying in the wind.

'Take a seat.' Johnson had climbed on to the board and was beckoning me to follow. I climbed up clumsily and nearly tipped him off. When I finally managed to sit down, the seat went on swaying for quite a while. My body only gradually adapted itself to the motion. My host did not need to explain anything. I began to understand, and the understanding came through my muscles and my sense of balance. Here was the simplest kind of equipment: it had reacted to me and forced me to react in turn. Meanwhile the swimmer had climbed out of the water and joined us on our swaying platform. It was Brodey.

He immediately began to lecture: 'Our environment and our products are hard, dumb and soulless; and so we too have become hard, closed-in and insensitive. Even an infant should lie on a living mattress with which it can communicate, on which it can try out and develop its faculties. We are thinking about chairs which will adapt to the sitter and possibly teach him new ways of sitting; about cups that will mould themselves to your lips and be agreeable to the touch; about shoes that will adapt to individual feet. Why do we have the same knives and forks for so many different hands, or vehicles which are rigid as tanks instead of being flexible, responsive and intelligent? Why shouldn't a sensitive car refuse to move if the driver is tight? I have built a caterpillar from plastic and electronic sensors which gives some idea of what an adaptable vehicle will be.'

'But don't you think people have other things to worry about, more important than getting slightly more human, slightly more agreeable equipment?'

'It's not a question of convenience,' Brodey replied quietly, almost gently, as he pressed a little wooden toy into my hand. It was like a piano key, and I noticed that it began to move in my grasp with a definite rhythm transferred to my hand from Brodey's by means of a thin wire.

'Answer,' he said.

I too stroked my key and knew that he felt the signal immediately.

'There,' he said. 'There you are.'

'What do you mean?'

'We must learn that, for our environment, it is we who are the environment. Our environment feels us just as you felt me and I felt you. We must learn that we are always affecting our environment in ways we do not notice or notice much too late because we have not developed the feeling for it. I want to tell the fish about the water – about their environment which they forget because they are completely enclosed in it. Partly because the outside world with all its dangers is soulless and silent. If only the push-buttons on the technological systems with which we can blow up our planet could challenge us, could warn us! They don't. They can't. We have become the slave-masters corrupted by our silent slaves.'

The attempt to bridge the chasm between man and his technological creations is among the most important tasks of our age. The effect of a rigid, insensitive, unsubtle, dangerous technology upon those who operate it is to make them rigid, unfeeling, coarse and dangerous too. Most of the work published today about the relationship of man and machine concentrates on the effect of technology on man and pays far too little attention to the possible influence of man on technology. Our ancestors did not invent it in their own image, but as a crutch for their – and our – physical weaknesses. We have become as dependent as invalids.

We cannot simply throw away these crutches, because they have changed man collectively and individually far more profoundly than the ascetics like to admit when they preach, understandably enough, that we should turn away from the world of machines which has begun to dominate us. But though we must accept technology in principle, does this mean that we must be resigned to it in its present form which underrates such specifically human faculties as feeling, intelligence and imagination? Certainly not.

Walter Rosenblith, Marvin Minsky and S. A. Papert are continuing the work of their teachers Norbert Wiener and Warren McCulloch at MIT by experimenting with the invention and development of equipment with built-in 'human' faculties such as perception, intelligence, sensitivity and flexibility. Similar work on evolutionary technology and artificial intelligence is being done in Europe and Asia. These things may strike today's observer as mere toys: in the Artificial Intelligence Group at MIT, for instance, robots pick up different coloured blocks and build primitive structures with them; Dewan and Farley have 'brain currents' that switch lights on and off; large numbers of research workers in the US, the Soviet Union, Britain, France and West Germany are developing robot systems which will be able to function on other planets and perform certain tasks with a degree of independence.

All these experiments produce ever more sophisticated micro-circuits with an ever closer resemblance to living brain cells in terms of size, sensitivity and speed: one day these things could alter man's relationship to his technological environment. The effects on those operating this equipment can only be guessed at. But among other things it is possible that man, by working with artificial creatures closer to his own level than the machinery of today, will come to a clearer understanding of his own unique faculties, the faculties he will never be able to imitate: joy and sadness, pity, love, dreaming and abstract thought.

Measuring the quality of life

These specifically human faculties – which are therefore felt as anthropological needs – demand other values and goals than those which have taken precedence so far this century. The intensive search for new horizons and new directions which is expressed in the slogan 'quality of life' goes far beyond the attempt to make a life-enhancing environment. After a long interval of suppression, the great questions about the meaning of life, earthly happiness and man's purpose are once more becoming our guiding principles.

The concepts of controlled, soft and evolutionary technology contribute to the greater importance of the part now played by non-economic factors. From now on, technological equipment and systems will be judged not only in terms of output: we shall also ask whether they are bad for health, whether they harm the environment, whether they contribute to the alienation of their operators. Such questions presuppose guiding values which need to be understood and measured if they are to be given their proper place in future planning.

Considerations of this kind have led to the idea that we need a system for measuring not only scientific and economic data but also the quality of life. According to this new scale of values, many politically important cities and industrial centres would lose the place they now occupy at the head of the economic scale and sink to middle or bottom positions, for such places are often significantly poor in terms of health, safety, peace and beauty. On the other hand, communities with fewer people, clean industry, no unemployment, good educational facilities, a lively cultural life and a varied environment would be rated much higher than they are today.

European efforts to judge progress and well-being not only in terms of production statistics go back to the nineteenth century, and interest in such atypical attitudes has never quite ceased. All the same, little attention was paid at first to books such as Alfredo Nicoforo's *La Misura della Vita* (The measure of life), which appeared in 1919, or to the pioneering work of the economist Karl W. Kapp in the thirties and forties, in which he tried to publicize 'the social cost of private ownership'. It was not until the sixties that the work of a few American political scientists and sociologists, and particularly that of the Englishman Edward Mishan, began to find a response: what they emphasized was the need for social as well as economic accounting.

I remember the occasion when the new term 'social indicators' was first used. It was in 1966, and Raymond Bauer, a social psychologist from Harvard, was addressing a conference on the future of the environment at Endicott House. He said that social indicators were necessary to measure the quality of life, and reported on the work of a government commission which had been sitting since March of that year and had been considering the following seven questions:

◇ What is our state of health?
◇ How much equality of opportunity is there?
◇ How is our environment?
◇ What are the job opportunities?
◇ What are the developments in crime and public safety?
◇ What about education, science and the arts?
◇ How much co-determination is there, and how much alienation?

The commission had been set up by President Johnson towards the end of his term of office. Its thirty-one members published a report saying that facts such as these about the state of the nation, although difficult to assess, should be collected and published each year.

Since then efforts have been made not only in the US but the whole world over to use other, not purely economic values as guidelines for a more humane future development. But if we consider these social indicators to be just as important as economic factors – perhaps more important still – then we must stop dismissing as 'uneconomic' or 'unproductive' the building and equipment of schools, hospitals, recreation centres, fast and comfortable public transport, factories harmless to the environment and machines suited to man. All these things are necessary to restore the lost quality of life, and if we recognize this the new technology will stand a chance.

Experiments, prototypes and new inventions whose aim is to create a clean, quiet, healthy, safe and stress-free technology will have no real chance unless they are supported by private or public funds; otherwise they will be unable to compete with ordinary technology which is cheaper and – from a short-term point of view – more efficient. Taking the long view, however, the new technology subservient to man will actually be cheaper, for its social cost will be much less. In the future we shall have to spend billions on undoing the damage which our present type of technology inflicts on man and his environment, not to speak of suffering irreparable and incalculable loss of life and variety; instead we could use more sensitive and less violent tools to make ourselves not only richer but happier as well.

In the technology of health, for instance, there are a number of innovations which are not being used simply because there is no profitable market for them. The same applies in the fields of accident prevention and of anti-noise, anti-smoke and anti-pollution measures.

There is no doubt that the development of a neotechnology to make life more bearable would require enormous sums of money at first. Some of this could be saved by disarmament. Seymour Melman of Columbia University, New York, has shown that if even a third of the American armaments budget were diverted to improving human conditions by technology, the benefits would be felt by every citizen.

But who will decide which of the projects dictated by social indicators shall get the money?

François Hetman, a French delegate to the OECD, believes that during the period of changeover a fourth power will be needed in addition to the existing powers of parliament, government and the law. This will be the 'innovating power' whose chief function will be to explore and invent new types of future. He proposes that each country should set up a 'National Council for Innovation' whose main task would be the democratic discussion of new technical developments and their order of priority. Who will be fitted to speak in such an assembly? Whose interests will be represented? How can we make sure that today's technocratic positions of power are not simply strengthened and extended?

Dr James B. Sullivan of the Center for Science in the Public Interest has been involved in attempts to have a say in technology assessment. He expresses his scepticism as follows: 'The most common response from a technology's developer to citizens seeking information is that they don't need it. Technical experts cost more than citizen groups can pay. While agency officials are required to hold hearings for citizens, they are not required to listen.'

New institutions like Hetman's Council for Innovation require a type of person rarely found today: the generalist. He is the opposite of the specialist: he has many interests, many sources of information, and the ability not to lose sight of the whole among the details. He must be open to the unexpected and have considerable mental flexibility. Even today such people are much sought after to fill positions in government and economics: they need to be able to combine knowledge from many different fields such as nuclear physics, the law, social anthropology, philosophy, biology and management.

The new technology will require new men.

4 More Democracy

Revitalizing politics

As I write these lines, the idea that democracy could be fun seems as unlikely as it is inappropriate. We have learnt to look upon the exercise of our political rights as a boring if solemn burden which we would prefer to avoid, just like school.

But this comparison is no longer wholly valid, for some modern teaching methods have succeeded in making many young people enthusiastic about going to school by awakening their pleasure in learning and experience. In the same way, the new kind of politics whose beginnings are just appearing could turn participation and co-determination in public affairs into an interesting, even enjoyable occupation. The West German experimental political television programme 'Oracle' was considered by one viewer 'as exciting as a football match' – a remark which shows the trend things might take. The aim of this kind of development would be to eliminate the alienation that has been growing for decades between the individual and the community. The attempt could succeed:

◊   if the gulf could be bridged that has opened up between citizens and political institutions because of the increase in population and the centralization of decision-making;

◊   if the complexity and confusion of politics were minimized by an open flow of information;

◊   if people affected by any public matter were able to voice their ideas and objections in time, and to be heard.

This revitalization of democracy could be assisted by electronic communications techniques which can bring distant objects close and unite partners in discussion however far apart they may be. Information could be quickly disseminated, it would be easily accessible, and – perhaps most important of all – enormous masses of confusing data could be clarified by emphasizing the most important elements and establishing meaningful connections. In this way the principle of democracy would cease to be an empty promise and become a reality. But here too the positive potentialities of the media can turn negative unless people develop to match the new techniques.

A 'human revolution' seeking to activate the imagination and to awaken the latent potential in every individual would be able to create by the end of this century the conditions which are necessary if democracy is to become more than just a catchword. It is only now that the masses are developing that courage to judge and criticize which earlier revolutions mistakenly took for granted. For a long time people underestimated how much dependence, obedience, ruling and giving orders had become habits over the centuries. Democracy, therefore, is only just beginning.

It is quite possible that the 'fear of being free' will remain dominant during the crises we must expect at the turn of the millennium. The majority of citizens are so intimidated by educational and power élites that they still consider themselves insufficiently qualified or incompetent to face the coming perils. But there is no ineluctable necessity for abdication in favour of the specialists and their back-room boys. The many signs of unrest in every country and continent go to prove it, and this unrest is more and more frequently accompanied by attempts to create and practise new democratic initiatives and to set up new democratic institutions.

So we are living through the renewal of democracy. As the huge political and economic systems become more rigid, so their petrifying shells contain more social developments whose aim is to activate frustrated democratic demands. Among such developments are the following:

◇   agricultural and urban communes;
◇   consumer associations;
◇   community projects;
◇   shopping co-operatives;
◇   neighbourhood and parental inter-aid;
◇   popular dispensaries and hospitals;
◇   building co-operatives;
◇   underground publishing, press, television;
◇   non-hierarchical production units owned by the producers.

The Englishman Stan Windass, the founder and leader of the Alternative Society group, keeps a register with the names and addresses of hundreds of attempts such as these at creating new democracies; so does Source Collective in Washington; and so does the magazine *Modern Utopian*, published in San Francisco. It is true, however, that if you write to these addresses there is often no reply.

If you try to go and have a look, you may find yourself standing before a locked door or vacant lot. For the death-rate among these experiments, as I mentioned in a preceding chapter, is almost as high as the birth-rate. But the end mostly implies no more than a transmogrification. Another experiment will begin in another place and in a different shape or combination, and the participants regard this very instability as a sign of vitality.

'Sure, we quarrelled,' a young psychiatrist said to me. Together with a like-minded group, he had started a 'psycho-store' in a San Francisco street. For two years they had run this corner stall for people with neurotic disorders: people flocked to it, and they achieved demonstrable successes. 'But it doesn't matter,' he said. 'Steve thought we were only curing symptoms and that it would be better to pull out the evil by its economic roots. So he has gone into politics and is trying to teach his pals that they haven't been paying enough attention to the psychological demands and ills of society. Lennie has gone back to the university. He's teaching management science to future bosses and he is going to make them construct models of socially responsible businesses. Zelda and I are trying to put former drug addicts back on their feet. But we're not just taking away their stuff: we're giving them something else in exchange: responsibility, work with other people, a goal in life. Or anyway an inkling of those things. You don't suppose the five of us who started the psycho-store have got weaker? Not at all: we've multiplied and strengthened our effectiveness.'

Specialists go to the people

Outside a Swedish supermarket a man with blond hair and heavy-rimmed glasses is handing out leaflets warning people against certain foodstuffs which have been treated with harmful chemicals either to attract more buyers by a more pleasing appearance or to increase their durability and save the shopkeeper losses.

But this activist not only distributes printed matter; he talks as well. 'You don't know what you're buying,' he tells people. 'That's just about understandable. But you don't even know who governs you, and that is really dangerous. There are groups in the government commission on the environment where representatives of the consumer industries are in the majority. They see to it that we go on buying washing powders that contaminate our rivers. Do you realize that? What are you doing about it?'

People crowd round the speaker, not only because what he says seems sound and sensible, but also because they want to talk to a celebrity. For Björn Gillberg, a geneticist, is not only one of the most controversial scientists in Sweden, but also one of the best-known – ever since he was deprived of his facilities for working at the microbiological institute of the agricultural college at Uppsala, presumably because he had publicly criticized the government for failing to control the use of dangerous additives. His criticisms implied that the government was ceding to pressure from business interests.

Of course, the authorities maintained that they had withheld funds for his research on a new type of organic manure for purely practical reasons, but the public would not buy that story. There was a spontaneous movement to collect money. Thousands of contributions were received from individual private citizens as well as from trade unions to enable Gillberg to continue his work at a new independent institute.

Gillberg, who has published internationally recognized work about the possible long-term ill effects of chemicals on human genetic material, judges his 'case' in a more general, social context: 'People today are more suspicious than they were four or five years ago. They have become more critical. If I have contributed to this development, then I'm satisfied.'

He has far-reaching goals. He wants to create a permanent alliance between scientists and ordinary citizens. Scientists should not publish only in specialist journals: they should give up their isolation, go among the people, and together with them fight the dangers that threaten us all, even if it means imperilling their jobs. Gillberg told the English scientific journalist Andrew Jamison: 'We need a system in which people have more influence.' He feels that we must work for a political system geared to the common man; the goal should be a much more personal style than heretofore.

There are hundreds of Gillbergs today, possibly thousands. There are the many young doctors, laboratory assistants and nurses who do voluntary work as 'advocate doctors' in the worst American slums. They are not content, like Samaritans of the past, to provide medical help: they enlighten people about their social position and explain to them why they fall ill so often and why many of them die young. Thus they encourage their patients to demand from their local authorities the rights and funds which are their due. They decline to behave like 'little Albert Schweitzers' and to adopt charismatic leader-like attitudes. Their aim is to get the people concerned enough to look after their own interests as far as possible, to manage their free clinics themselves and to receive only help and advice.

I saw some of these 'people's medical centres' in Chicago's 'Bronzeville' and the Puerto Rican barrio of New York. They are no longer run like military units with white-coated generals, colonels and lieutenants and severe non-commissioned officers: the atmosphere is open, fraternal and cheerful in spite of the physical misery. I watched an old man being examined for severe abdominal pains. The doctor was externally scarcely distinguishable from the patients, and he thought out loud. Before each step in his examination he told the patient exactly what he was going to do and what he hoped to discover from it. There was no learned secrecy, no trace of magical authority, but instead an effort to make the patient understand and participate.

The 'advocate planners' are amazingly similar in their behaviour. I got to know them in October 1967 on the occasion of a meeting to celebrate the fiftieth jubilee of the American Institute of Planning in Washington. A counter-conference was held in the garden of the very hotel in which the official congress was meeting. The official programme included visits to admire new model housing estates, administrative buildings and office blocks; the counter-programme organized tours of the slums of the nation's capital, of open spaces that had been built over, of new buildings without sound-proofing and already beginning to look dilapidated, of speculative building projects undertaken without proper permission. We heard of extortionate estate agents, and talked to families who had lived for years under threat of eviction.

They told us how the 'advocate planners' and their energetic and passionate originator Paul Davidoff had not only tried to help them in word and deed, but had also encouraged them to discover and voice their own complaints and wishes and to take political action accordingly. The legal expertise of the 'action lawyers' was of the greatest importance; they were able to use the machinery of justice skilfully to prevent evictions, as well as to formulate applications for public funds and even to initiate legislation.

We can take Professor John Coons of the University of California at Berkeley as the prototype of these committed lawyers. He and his students Steve Sugarman and Bill Clune worked painstakingly for years to mobilize legal experts and officials into bringing a case against the unfair distribution of funds between well-to-do middle-class schools on the one hand and ghetto schools on the other. This case became a model for many others.

The emergence of a younger generation of socially responsible lawyers has become known to the world at large above all through the person of Ralph Nader, the son of Lebanese immigrants, who started the consumer movement. To name only a few of the campaigns that Nader fought successfully with the help of his expert assistants, there was the apparently hopeless but eventually triumphant battle with the giant General Motors Company; the revelation of mismanagement in hospital administration; the crusade for clean air carried out with a staff of volunteer students and lawyers; the fight for the neglected rights of Vietnam veterans; the ruthless investigations of such sacred institutions as the New York banks, the House of Representatives, the National Science Foundation. All these undertakings showed how impenetrable to the ordinary citizen the economic structure and the government institutions of the US had become. The campaigns for the enlightenment of the consumer were effective beyond their immediate goals: they proved to an indifferent public how important it was to have more publicity, more say and more participation for the citizens if they were not to be constantly disadvantaged and robbed of their rightful claims. Nader has understood better than anyone else how to mobilize students for his attempts to effect urgent changes. He won several hundred thousand dollars in a claim against General Motors on an invasion of privacy charge, and from 1970 onward he used this money to set up the Center for Responsive Law in Washington D.C. and Public Interest Response Groups (PIRG) at several American State universities. Bo Burlingham has described how these groups function:

> Organizers would go out to college campuses and boost the concept of the PIRG. They would find committed students, who would organize referenda at their schools to have every student pay a small sum (3 to 4 dollars a year) towards financing the local PIRG. Students would then elect representatives to the PIRG's board of directors. The board would use the money to hire a professional staff (mostly lawyers, but also organizers, scientists, doctors, scholars and the like). . . . The PIRGs have succeeded in generating publicity on issues ranging from the auto repair racket (in Oregon) to a law barring 19 year olds from seeking electoral office (in Minnesota). In Vermont the PIRG has worked with the Vermont Alliance – a grassroot citizen action group – to force power companies to equalize rates now heavily weighted in favor of

large industrial users. Also, it has teamed up with other groups in Vermont to pass a bill providing for the first statefunded dental care programme in America.

Increased contact between qualified members of the professions and wider sections of society has had one important subsidiary result: the gradual growth of a common language. It is hard for intellectuals to express themselves so that they can be universally understood, because ordinary words often cannot convey exactly what can be said in specialist terms. Language barriers are erected not only from intellectual conceit, but also because it is often difficult to be accurate and easily comprehensible at the same time. Sometimes only experienced 'translators' such as science writers can cope with this dilemma. They have to compromise by sacrificing a certain amount of precision – but such sacrifice is often more acceptable than pure scientists will admit.

The problem of communication is becoming increasingly difficult, not only between the élite and the general public, but also – owing to the increase in specialization – between different groups of experts and managers. The above-mentioned unsystematic efforts at interpretation by a few pioneers of democratization are no longer sufficient. In future every trained expert will have to learn how to explain his knowledge and his skills to those who are not specialists in his field. It will therefore be important to provide courses in 'public language' at all colleges and universities. Perhaps the Chinese method might prove helpful: in China all students are required to spend part of their time in agriculture or industry.

The English are pre-eminent at expressing complicated scientific and technical facts in an uncomplicated manner. It cannot be only because of the loose and flexible nature of their language, otherwise American scientists would be equally able to express themselves simply and comprehensibly. In my view it is due to the special character of English academics, whose relaxed and modest manner is conspicuous at every international gathering. The breaking down of speech barriers seems to depend very much upon the human attitude behind them: we should try to eliminate attempts to impress, intellectual pride and linguistic pedantry.

The liberation of information

An essential problem – if not the central problem – in strengthening democracy is the question of the freedom of information. It is a two-way problem and consists of both the freedom to communicate and the freedom to gather important information. Yet we are always being told that there is far too much information, that we are being suffocated with data and therefore cannot orient ourselves.

But in this generalized form that is simply not true. It is true that the absolute quantity of information of all kinds increases all the time, but its meaningful content diminishes. It is the most important, the most consequential facts that are most often suppressed. This is so not only for political and military matters, for new inventions and products in the planning stage, but also for so-called harmless questions that concern people personally, such as town planning or labour-saving changes and innovations in the production process with consequent major effects upon employment.

Once you penetrate into the world of 'inside knowledge', you see everything with different eyes. I remember a conversation with Egon Bahr, the German Cabinet Minister and friend of Willy Brandt, who had worked for many years as a journalist in current affairs; he told a small group of friends what an experience it had been for him when he found himself in the Foreign Ministry of the Bonn coalition government and began to see the multitude of diplomatic and secret service reports after being used to receiving only incomplete and filtered information. He added that now he could scarcely imagine life without this access to proprietary knowledge.

I have always heard this experience confirmed by people with access to secret information. Even those who pride themselves on being politically well informed are allowed to hear and read only a fraction of what goes on. Some of it they discover years afterwards as a belated revelation, a piece of ancient history.

People have asked themselves how Daniel Ellsberg could have remained so long in the RAND Corporation working for the government even after he had become convinced of the wickedness of US policy in Vietnam and was accusing himself of being a war criminal. He told me about it in a long conversation on the beach at Malibu. At the time he was already on leave from the think tank because of his first letter to the *New York Times* protesting against Vietnam policies. He was sticking it out, he said, because he felt he ought to remain as long as possible in a position where he was one of the few Americans with 'the right to know', the right to read secret reports hidden from almost every other citizen. It is well known that later Ellsberg temporarily broke down the wall of secrecy by clandestinely copying the Pentagon Papers and distributing them to leading newspapers. But that put him finally outside the pale and prevented him from being able ever again to judge dangerous developments correctly on the basis of accurate knowledge of the true facts.

Another example, this time from industry: Arthur Kramish, a long-standing member of leading think tanks (such as RAND, the Stanford Research Institute and the Institute for the Future) decided to give up his career of prognostication because he realized that it was impossible to base prognosis on reliable data so long as industry kept back as 'proprietory information' facts about the harmful side-effects of their work as well as about decisive new discoveries.

In the last decade increasing secrecy about information has led several industrial nations to set up private espionage organizations, usually manned by experienced former employees of government intelligence departments. Like the infamous 'Intertel', for instance, a private organization based mainly on New York City, they usually work for big industry. On the other hand, there is also an increase in groups trying to uncover dangerous developments and plans in industrial and government establishments in the interests of the public. They are frequently able to use information from 'inside', which is passed on anonymously by 'whistle-blowers' who are disturbed by what is happening. Most of these 'citizen spies' maintain their incognito. An exception to this rule are the 'Spies for Peace' in Britain, who feed the press with carefully guarded state secrets.

Certainly the public all over the world is becoming increasingly dissatisfied with the fact that more and more information is being withheld from all but those with special privileges: such facts as local plans to build over public parks, regional plans for new airports, and even national and international planning such as strategic decisions. Since information of a kind has begun to be made available, the liveliest minds can no longer rest content with being allowed to see only the outward façade of what goes on. They want more: they want to know everything they need to know that could be significant for them or for the community.

It is becoming more and more common for officials and white- and blue-collar workers secretly to pass on facts that they have learnt in the course of their work about processes that might be dangerous to the public. They do this in protest against the secrecy of the firms or authorities who employ them. In this way an extensive and detailed underground information system, both written and oral, is coming into existence. It will doubtless gain in importance owing to the growing sense of social responsibility among the younger generation of workers. It could develop into a sharp minor war between the 'security forces' on the one hand and the employees on the other. But to take a more optimistic view, it could also lead to a gradual breakdown of the system of privileged secrecy.

There is another form of discontent that could have serious consequences among those sections of society which used to be deprived of information but are now receiving it through the mass media. Listeners and especially viewers are growing tired of their silent role. They want to react to the mass of images and information that floods in upon them all day long. 'Talk back to your television' is a slogan invented by Nicholas Johnson, a member of the US television council. His former colleagues on the Federal Communications Commission were all loyal government conformists who regarded him as a rebel with the public interest far too much at heart.

How could his slogan be put into practice? How can a free and open communications era become a reality – an era when, according to the social economist Robert Theobald, 'information will be more important than money'? Theobald is the apostle of the communications era. He is an Englishman born in India who now lives in the States. He began as a university teacher, but abandoned the academic milieu first of Cambridge, England, and then of Cambridge, Massachusetts. He recognized that in an age of telecommunications large cities are not the only possible information centres, so he abandoned the frantic atmosphere of New York as well, and set up shop in an old school house in the little town of Wickenburg in Arizona. From this centre he strives unceasingly to persuade as many of his fellow-men as he can reach to take an active part in the great changes of the millennium.

He thinks that our present capacity for gathering, storing and reproducing information could help in the elimination of poverty on a world scale, but that this possibility of satisfying the needs of mankind can only be realized in a society which considers new ideas critically and openly. We are in transition, he says, from an order based on commerce and production to one relying on communications, and our most urgent need is for a new language and new symbols which will replace power over people with the power of the people. We must design a world where development and progress is achieved by humans working together rather than humans supervising one another.

But Theobald is not the man to let these ideas remain pious wishes. He realizes them in various ways. For years he has been increasingly successful at stimulating the formation of groups throughout the US in which spiritual and material problems of the day can be discussed in a spirit of self-criticism. Initiatives, projects and experiments that result from these meetings are published in the monthly newsletter *Futures Conditional*. Its readers are specifically encouraged to 'carry on the creative process' by criticizing or collaborating. Under the heading 'Opportunities for Involvement' you will find entries like the following:

HEALTH. A group has been formed to examine how to create patterns of promotive health care. One meeting has been held and another is planned for the Spring. Contact: The Strodes, 1629 Wilder # 501, Honolulu, Hawaii 96822. Tel. 806-949-3840.

COMMUNITY-BASED LEARNING. An experimental environmental education program has been developed around the real problems of the Kewaunee River Basin, with close non-hierarchical relationships between students, teachers and community. Contact: Tom Abels, University of Wisconsin at Green Bay, Green Bay, Wisconsin 54302.

A DICTIONARY FOR THE COMMUNICATIONS ERA. Different language is developing today to ensure greater clarity in the communications era. A dictionary of the new language is planned. 'Want to help? Pick a word, or several words, that have particular meanings for you and try to define them. Use examples, sources, stories, songs, quotations, whatever makes the word clear.' Contact: Sharon Lynn, 6741 Fairfax Road # 21, Chevy Chase, Maryland 20015. Tel. 301-652-8450.

THE UTOPIA GAME. This semi-structured game for creating the future is being played for the first time at the University of Michigan. Its inventor wants help and feedback to make it more effective. Contact: Ken Davis, 1120 McIntyre, Ann Arbor, Michigan 48105. Tel. 313-761-7851.

Theobald realizes that the numerous groups he has encouraged or begun can only become effective if they unite to form an information network. He knows that ever since 1968 a number of similar networks have sprung up in Europe and the US with the purpose of working towards the future.

He wants to keep the loose structure among the groups exchanging information and mutual assistance. More rigid party-like structures would probably soon lead to the establishment of power centres and programmes. This would mean a loss of openness and flexibility. Theobald hopes that in the communications era movements which are geographically far apart will be held together by post, by telephone and by travel without losing their spontaneity. But will that be enough? In the end new institutions will have to be established to enable citizens to influence the political process with their own initiatives for renewal.

The democracy of participation

The invention and development of just and humane democratic institutions will be one of man's chief tasks at the turn of the millennium. In the coming years well-informed and public-spirited personalities will come together more and more frequently to discuss how the democracy of acclamation can be transformed into a democracy of participation.

A group of this kind was formed in Switzerland at the beginning of the seventies. It consists of members of the professional middle classes aged between thirty and forty who were critical of the establishment but not prepared to condemn bourgeois democracy altogether. Their conversations originated in a controversy that had broken out all over Switzerland after a questionnaire about the future of the country and its constitution had been circulated by members of parliament. A radical attitude was adopted with a view to redefining existing values as well as economic and governmental structures in the light of present anxieties and future problems.

Rudolf Schilling, one of the initiators of the working group, 'Swiss Alternatives', defines one of the most important results of their discussions as follows: 'The whole political issue and public discussion revolves around the question how. We do not need to ask why.' He points out one of the chief weaknesses of modern democracy: even in Switzerland, that model of democracy with its frequent direct referenda, the citizens are not admitted to the decision-making process until far too late. The definition of goals and the formulation of projects take place in a small circle to whose discussions the public has no access. Its opinion is not asked – if it is asked at all – until shortly before the final decision. The citizen has no influence on long-term planning; usually he does not even know about it.

But Schilling demands that the authorities should apprise themselves of the citizens' opinions in good enough time to act upon them. With the help of an extended system of communication they should plan and rule *with* the citizens. And the citizens should be made aware that they can participate in government.

Something like this was attempted at Baden, a medium-sized town near Zürich. Citizens' meetings were arranged to discuss the broad outlines of new plans long before they were put into practice. First they voted only on the town council's declaration of intentions. It was not a formal proposal for legislation, but contained merely principles and directives. Later the citizens were presented with detailed proposals for concrete measures. Later still, there were public hearings at which individuals could express their views, as well as opinion polls, orientation meetings and a quite unusual degree of openness towards the press and broadcasting media. As a result the Baden model became a much discussed prototype.

In 1969 an even more advanced plan for public participation in community planning was published by a British government committee, the Committee on Public Participation in Planning. It advised local authorities to set up permanent forums for planning matters. The idea was that different active groups and individuals should establish relations among themselves as well as with the local authorities. The expense was to fall on the citizens themselves. The regional authorities were to make only small contributions, which would give them no right of control.

A citizen's forum has the following duties:

◇ It creates opportunities for discussion, organizes exhibitions and film shows.

◇ It collects and collates opinions from individuals and groups.

◇ It provides or collects information data.

◇ It takes the initiative in advising individual groups when the time has come to take a stand on issues about to be discussed in parliamentary committee.

An especially interesting proposal is that a forum should have at least one 'development officer' who would be a sort of institutionalized advocate planner. He would help those to express their opinions 'who are not sufficiently educated, who cannot say what they want, who have never learnt to express their doubts, criticism, or wishes'. In the same connection, the American Donald Schon speaks of the necessity for 'social changers' who would mediate and help people in situations of social change, and enable them not only to accept alterations, but to understand and participate in them.

In tomorrow's polling booths the voters will no longer be treated like illiterates as they are today, when they are allowed to express themselves only by making a cross, pressing a button or pulling a lever. Perhaps the voter will find a teleprinter on which he can give a detailed opinion on the matter to be decided by the election. Work is proceeding on a recording machine linked to a computer which will not only collect the opinion of every voter but will summarize and evaluate the opinions with the help of a retrieval system. Voting results will then no longer be mere columns of figures, but will show a differentiated opinion profile which will give a much clearer picture of the citizens' opinions than we get today.

For the present, voting machines are not yet able to encode or decode verbal statements more complex than a simple yes or no. But there is no doubt that this technical problem can be solved and that, if he prefers, the citizen will not even have to write down his wishes but will simply be able to speak them into a microphone or telephone receiver. Even now it is technically possible to abstract key words from such statements and to summarize them. For instance, in the West German elections following the debates on the *Ostpolitik* the voters would have been able to say whether they preferred the draft treaty as it stood, or whether they wanted it strengthened, widened, weakened or rejected – thereby influencing the further course of events by their judgment.

Thomas B. Sheridan, Professor of Mechanical Engineering at MIT, deals in his chief lecture with the theme 'Technology, Values and Social Choice': like many innovators he is oriented towards interdisciplinary studies and combines psychology with his technical interests. He has set himself the following problem: radio and television provide governments with ever growing facilities for informing their citizens of their opinions 'from above'; how can we devise a technical process whereby the citizens can answer back? Sheridan's polling booth has the advantage of being technically immediately practicable: the voter finds himself at a desk with ten switches that can be set at yes or no, each switch corresponding to a given aspect of the problem under discussion.

He explains that the ten switches (if one takes all the combinations into consideration) offer 2^{10} alternatives – 1024. That is obviously too many possible answers to a single question. But the question could be posed in ten parts, each of which offered two alternatives, or in five parts, each with four possible answers, and so on.

The voter's wishes could be expressed even more precisely if he were able to weight his answers according to the importance he attached to the questions. For instance, each voter could have one hundred points which he could allot according to his views. Questions that did not interest him would remain unanswered, and he could use the points saved to express his opinion particularly forcefully on questions which seemed to him especially controversial. This kind of weighting could be built into Sheridan's machine in the form of an intensity switch.

But does the average citizen even know enough about the questions being put to the vote? Several investigations suggest that only a proportion of voters do. In his MIT Community Dialogue Project of 1972, Professor Sheridan produced a simplified form of his voting machine with the object of making citizens better informed on the problems for debate. Each participant – there can be up to ninety – holds a small ten-position thumb-wheel switch connected to an electronic display which is visible to all. The moderator or one of the participants asks a question and a short discussion of it leads to the establishment of between two and ten possible answers. There is then an anonymous vote, and the results form the basis for more detailed discussion.

An even more convincing form of intellectual preparation for voting has been suggested by Stuart Umpleby of the University of Illinois at Urbana. He wants to use teaching laboratories to improve and democratize political decision-making. Such laboratories already exist in many countries and will presumably become much more widespread. It is well known that with these teaching machines the student can project textual, graphic and other optical types of information on a screen, working at his own speed and in as much detail as he thinks necessary.

It was in 1965 in his laboratory at Urbana that Umpleby's teacher, the linguist Charles Osgood, first showed me how a teaching machine could be used to assess problems in futurology. He had developed the PLATO programme to store the various alternative developments that could be foreseen at the time. If the questioner chose, for example, the problem 'motor traffic', he could then follow through the alternative possibilities step by step – more individual cars, or the improvement of public transport, or a variety of intermediate solutions – each with its probable consequences. Umpleby and his collaborator Valerie Lamont thought this machine might be ideal to help voters make better decisions on the basis of detailed individual information.

The first experiment in the field was undertaken in February 1971. The problem was to establish how the citizens of Urbana would react to the planned regulation of a badly polluted stream that ran through the town. Should it be built over and turned into a sewer? Should all industrial building be banished, the stream purified and its banks landscaped? Or would it be better, in view of the shortage of public funds, to leave everything as it was? In addition, there were a number of compromise proposals, and, naturally, technical, hygienic, psychological and financial factors had to be considered.

Input machines were then set up in several schools and public buildings. They were connected to the computer by cable, and those taking part in the opinion poll were able to get from them all the necessary data; they could also play over on screens the probable consequences of various alternative decisions.

The results of this experiment were exceedingly encouraging. Most of the participants admitted that they had been better informed than ever before about the probable consequences of their decisions. They liked being instructed as well as questioned, and, with a very few exceptions, they never suspected that the programmers might have fed the computer with biased data.

This was a 'citizen sample simulation' with comparatively few (130) selected participants. Similar experiments, with an increasing number of input machines (up to four thousand), have been conducted in Urbana. They were to help decide whether the process should be used for voters in all local elections. Unfortunately, this promising experiment was abandoned, like so many others, for lack of interest on the part of the authorities.

But with most projects for using the latest information techniques to improve democratic participation, one must consider how the executive is going to cope with the vast increase of information coming 'from below'. Various alternative solutions were tried out over several years in Puerto Rico after the New Progressive Party came to power and its leader, Louis A. Ferré, became governor. He had studied in the US, and sent for one of his former professors, Dr Chandler Harrison Stevens, to speed up the democratization which the island had been promised under the slogan *Nueva Vida* (new life). This was the start of the PRIDE programme. PRIDE is one of the usual mnemonic abbreviations for the name of an organization, in this case the Puerto Rican Information and Decision Environment.

The centre of the programme was the communications room in the Governor's Palace. On a smaller scale it resembled the command centre of the Space Authority at Houston: there were data machines for storing and retrieving information, and large screens on which computer diagrams, statistics, incoming news bulletins, plans of all kinds and occasionally television pictures from one of the twenty bureaux for citizens' aid could be projected. All over the country bureaux were open twelve hours a day where all adult Puerto Ricans could register their complaints, wishes and suggestions. Significantly, these bureaux were housed in the fire stations. Alarm bells would ring if something was wrong in the realm of the people's king Louis. His social firemen, the 'citizen aides', were trained as generals in every conceivable field.

In fact these alarm posts were simply information bureaux where friendly young people were prepared to answer questions from a 'public service handbook' which they themselves had helped to compile. Their job was to get the Puerto Ricans to talk openly, to take a share in public duties and to discuss the future of the country. But such things take time. People are needed who can express themselves without fear, who trust the authorities in spite of previous bad experiences, and who are prepared to risk the 'revolution in mutual understanding' which Governor Ferré was continually proclaiming.

'I shall make the first attempt to govern a country with information for and from everyone,' he declared. 'Our new institutions anticipate the education of our people and encourage peaceful change.' But as Ferré's defeat at the polls in 1972 and subsequent retirement showed, the gap between dream and reality was still too great; indeed there is evidence that the reformer wanted to establish an autocratic régime behind a façade of brand new democracy.

Political consciousness through people's television

In the first years after the Russian Revolution, most people thought that that huge agricultural country would never be able to build up its industrial strength because there were not enough trained workers. The money spent on foreign plant and machinery was therefore wasted. It would never be possible to get the mujiks to set up a complex technical system, let alone maintain or extend it.

Today, the same mistake is being made over investing in more democracy. We are told that all attempts to let the citizen participate in political and economic planning and decision-making processes are useless because the great masses are simply not interested and will never be able to understand complex situations that can only be grasped by trained specialists.

This prognosis will presumably be proved as mistaken as the pessimism about Russia's modernization. It underrates man's ability to learn. It is an expression of conceit and thoughtlessness, and blames the underprivileged for their backwardness, which is the result of the selfish behaviour of the upper and middle classes. The creation of industrial complexes in an agrarian country gave the impetus for the training of millions of skilled workers, specialists and inventors within a single generation; in exactly the same way, the anticipatory setting up of democratic institutions will reawaken people's atrophied interest in the fate of the community and enable millions of citizens to develop the knowledge and ability that will make them capable of responsible participation.

However, this will scarcely happen if 'political education' is dispensed in the usual manner by lecturers to bored audiences or party conferences. It has long been recognized in language teaching and physical training that quick and willing learning is the result of letting the students participate on a practical level. Political consciousness will develop much more slowly through long theoretical courses than through demonstrations, tenants' associations, community projects or the cross-questioning of politicians.

All these opportunities for democratic activity arise quite spontaneously. But they should be greatly increased. Since the invention of parliament in the West there have been hardly any innovations in the way of democratic institutions to keep pace with the huge increase in population, the changing way of life, the new technical facilities and the rising (though too slowly rising) level of education.

Such innovations are quickly accepted and not only give pleasure but also train the participants in self-confidence and public spirit. The Canadian *télévision communautaire* provides an example. This experiment came about by chance. In order to improve television reception in that vast country, TV cable firms set up large antennae, laid cables, and offered to connect anyone prepared to spend a few dollars a month. The viewer was then able to tune into up to twelve different programmes with excellent reception.

One of the *cableurs* in Montreal hit upon the novel idea of devoting an independent channel exclusively to time and weather information. This was the start of the highly popular Channel 9. The director reported that he found it rather boring in the long run to look at nothing but numbers and measuring instruments on the screen – hour hands, minute hands, barometers and so on. He wondered whether one could make it more interesting by means of local news or with the appearance of amateurs, people who could tell their unknown neighbours something about their work or their professions. He tried this, and the participants were delighted. Others followed this lead, and soon there were hundreds of such programmes. By this means Canadians no longer needed to be passive spectators, but could partly produce their own programmes.

The Ministry of Education began to be interested, and subsidized local channels which sometimes serve only a few thousand viewers.

Technical advisers and trained television personnel were made available, but they were to remain in the background as far as possible and to retire after a few weeks. Everyone is given a chance to work as cameraman, announcer, sound engineer, producer or actor. Even children are allowed into the studio and are beginning to make their own programmes. Adolescents are given portable video equipment and make critical films about their environment. If they turn out successfully, the films are not sent out just once: they are shown repeatedly, and discussed, in special local cinemas.

Very soon social effects and side-effects turned out to be the most important result of this innovation. The citizens with their cameras penetrated into the council chambers of local and regional government. Political meetings had gone out of fashion, but now they began to attract enormous attention since the participants realized that they might be visited by camera teams who would carry their deliberations into thousands of homes.

This in turn resulted in an increase in political programmes on national television. 'Sur le vif' (Taken from life) is a Canadian Broadcasting Corporation programme on which anyone can appear who has a complaint to make. Journalists follow up the complaints and try to investigate, clarify and help.

French television produced a programme called 'Twenty Million Citizen Cameras'; the subject was Canada's democratic renaissance as a result of this use of the new media. It was conceived by Pierre Schaeffer, the head of the experimental division of the ORTF (French television) and an extraordinary man – composer, novelist, essayist and creator of *musique concrète* – who for years has been trying to propel the 'blocked culture' of his country towards the future. He comments as follows on television's democratic possibilities:

> We should be able to use television like a telephone. People have to learn again to speak to each other. But we must also make it clear to the viewers that the pictures are only shadows, only a tool for discovering real daily life. The medium can teach them to discover people, groups, city life and country life. But unless something at least of themselves is added, nothing will remain but the passing on of someone else's knowledge, or the portrayal of power relationships which are being rejected. I hope that these pessimistic prospects will be outweighed by a real breakthrough of possibilities for free expression. I think this way would lead to a new culture – but one which would have to be quite different from the one we are still portraying today.

Video guerrillas

Among the oldest human customs are the mourning of the dead and the various funeral ceremonies. I experienced how the dead are honoured in the media age when Jerry S. invited me with other friends and acquaintances to an electronic vigil for the dead which he was holding in his apartment on Riverside Drive, New York. Every evening for a whole week the three darkened rooms were crowded with people watching the shadows of the late Anthony S. and his family flitting across several monitor screens.

The father had suffered from stomach cancer for the last three years of his life. During that time, the son had made innumerable videotapes of him: S. swallows his breakfast, S. kisses his family, S. gets into his car to go to the office. The car won't start, so S. has to get a taxi. S. visits his customers, has a row with them, tells jokes to help along the deal. S. in a crowded cafeteria. S. does his accounts, disturbed by the din from the street outside. S. looks at girls in the crowds on Fifth Avenue. S. slaps his little daughter, S. cuddles his little daughter, S. goes to sleep in front of the television set, S. plays with his family on a crowded beach, laughs, splashes up the water. S. lights a cigar, S reads the paper, S. grins from ear to ear.

All this is accompanied by the son's voice in the dark: 'He was already a dying man. He did not know it. Or he did not want to know it. He went on going to work. He was getting weaker. His bosses must not know. They would have sacked him. Finally he went to the doctor. But it was too expensive. He lied to us, and told us he was having treatment. Once I accompanied him to an out-patients' department. But there were dozens of people sitting there. Can you see them, all those doomed people? He lost patience and ran off. I begged him to wait. He ran away from me.'

You can see him running away on the screen. And the son's voice shouting in protest against the senselessness of life, against people's untruthfulness, against the lies of the advertisements, 'against all that shit around us'. Meanwhile the camera shows the filth of New York, the garbage bins, human wrecks, car wrecks in the streets, the slack faces on the subway, the blank eyes behind the tills, the open, swearing mouths of taxi drivers . . .

It goes on hour after hour. Innumerable yards of videotaped torment. Sometimes it is monotonous – when we see Anthony S. shaving or his wife shopping in the supermarket with S. trotting behind, or when the son trains his camera for four or five minutes on his father's sleeping face and no commentary interrupts the snores.

Lack of respect? No, he had wanted to honour his father by show-ing the truth about his life, said twenty-three-year-old Jerry S.: 'That's what being a video freak is all about. We want to see reality and make others see it. Tear down the veils and packaging of adver-tising and propaganda. All the illusions about our environment and ourselves. That's why we tape hours of conversation, and our most ordinary activities. Some even tape themselves making love. It isn't narcissism, as some people say. Narcissus was in love with his reflec-tion. We look at ourselves critically. Sometimes we hate it. But it makes us think. How we got like that. How we could change it.'

At the end of the sixties, a Japanese firm began to market the first portable video-magnetophone equipment with tapes that could be played back at once and wiped off for new takes; immediately dozens of video groups sprang up in the US with names like 'Televisionary Association', 'Experimental Video', 'Free America', 'People's Video', 'Community Video', 'Global Village', 'Earth Light', 'Media Access Center', 'Children's Television Workshop'.

Forty or fifty people would club together to buy a video recorder, which still costs more than $1,000, and start shooting their own films. The machine is so easy to handle that even small children can manage it with a few instructions: a Californian group at the Media Access Center at Palo Alto even specializes in working with four- to sixteen-year-olds. In a school with a lot of conflict between blacks and whites, video films began to build the first bridge of understanding. In a high school near San Francisco the children were given the video project 'Our School and Its Environment'. They included a sequence called 'Strange and Sometimes Dangerous Animals', which was a ruthless reportage on the teachers – their peculiarities, their tempers, their unfairness, their occasional attempts to appear benevolent or even friendly. After the performance the film was run off several times in the staff room where it was then decided – again without revealing the plan – to make an equally critical film about the children. As a result, relations between the generations were greatly improved.

The two best-known groups are 'Raindance' (an allusion to the RAND Corporation) and 'Video Freaks'. One of the founders of Raindance is Paul Reilly, who used to work for the media prophet Marshall McLuhan at Fordham University, New York. Once when he was spending a period in jail as a conscientious objector he de-cided to put his teacher's ideas about communications in the elec-tronic era into practice. When he got out he joined up with a young journalist from *Life* called Michael Shamberg, and together they made a video film of interviews with hippies, street singers and runaway children. The subject was the recently fashionable East Village on lower Second Avenue, New York, with its psychedelic pseudo-paradises, its little bookshops, marijuana parties and way-out theatrical performances.

The mass media usually presented the place as a romantic island in the midst of tough Manhattan, full of dreams, leisure and spontaneous gaiety. Reilly and Shamberg showed the other side: the hangovers, cold turkey, hunger, begging, pathetic little thefts. But the video interviews with acid heads, winos, sex maniacs, and freaks of every kind showed what these people were really searching for, what perfectly justifiable wishes they wanted to satisfy, and how much they suffered in the heartless commercial world of Manhattan.

I saw these and other similar tapes in a shabby, stuffy little public cinema next to a sauna bath on St Mark's Place. Frequently not just one film but three would be shown simultaneously on the same big screen. At first, if you are not used to it, this seems confusing, even senseless. But gradually you come to understand and even enjoy it. It is an attempt to escape from the narrow and explicit world of the written and printed word and to enlarge one's experience through a wealth of impressions. It helps to develop our far too rarely used faculty of absorbing several impressions simultaneously and subliminally.

According to Michael Shamberg, the video guerrillas' chief aim is to make it possible to show all viewpoints, particularly those which are not truthfully presented in the conventional media. This happens when people can give information about themselves without being hampered by any invisible controls.

This attempt to create a television underground is making significant progress on the North American continent. The TV guerrillas have a magazine in which the little video cinemas that have sprung up, especially in university towns, can announce their programmes. Media buses tour the country and are particularly popular in small places, not only because they provide uncensored reports on the big cities, but also because they give local inhabitants an opportunity to see themselves as actors on the screen presenting their local problems in their own unsophisticated language.

Ordinary television normally leads to isolation and retreat into the home; but the Canadian *télévision communautaire* and the work of the TV guerrillas encourages community life. 'Big television' leads to passivity, but 'little television' encourages active participation and prepares people for the growing opportunities of using their own channels with cable television – and if these channels remain unused they will soon be taken over by advertising and the entertainment industry.

The young Canadian researcher Jean Cloutier of the French University of Montreal has invented a prototype of the active participant in the communications process – a model for the generation growing up today. He calls him EMEREC (an abbreviation for *émitteur* = sender and *récepteur* = receiver) to show quite clearly that he will be no mere passive spectator and listener, but primarily an active communicator. Cloutier thinks that schools ought to teach the use of the new tools of communication (camera, microphone, TV stations) just as they teach pupils how to use pens and typewriters. His own son has been brought up in this way. 'At the age of nine he went out with his video pack to interview an airline pilot,' Cloutier told me. 'He thinks that's quite natural. If only the equipment were not so heavy. At the moment lugging it about is what he finds hardest.'

During the sixties a wide variety of attempts at audience participation was made within the big private and state television networks as well. In England Doreen Stevens tried to reproduce Hyde Park Speaker's Corner in the studio. Her programme was called 'Roundhouse' and was conducted in a completely natural, open tone such as had never been heard on British television before. But in Britain a person who thinks himself publicly libelled can sue for high damages, so the producers of London Weekend Television decided that this outspoken programme was too risky and abandoned the experiment.

Another British company, Thames Television, tried to create a platform for ideas and to encourage reforms. Working with the National Suggestions Centre, they produced a weekly programme called 'What?' Viewers were invited to send in ideas for necessary improvements. In three months more than a thousand suggestions had come in from Greater London alone. A few of these were later put into practice in co-operation with the Post Office: among them were a telephone information service on the daily market price of vegetables, and a considerable reduction of telephone costs for old people. In Norway a similar programme called 'Ideas Bank' led to several practical improvements and social reforms, especially in areas far removed from the capital.

One of the earliest and most successful attempts to instil life and interest into democratic decision-making processes came from the University of Iowa television station at Ames. At first it was difficult to break down the reticence of rural communities who were suddenly to have their council meetings exposed to public view. But the television crews were so friendly, unassuming and simple in their manner that they quickly succeeded in encouraging the participants. Soon these debates about local problems in the state of Iowa became so popular that viewers began to neglect more expensively produced entertainment programmes for their sake. After a programme made in a village of six hundred inhabitants, students reported as follows: 'When we entered Cambridge we found an apathetic, dispirited community, afraid to discuss problems. In the past few weeks we have watched a ferment growing in this town. We have watched people as they began to talk about their problems in the open – for the first time.'

In a final evaluation of their project the students wrote: 'The people themselves – not experts, but the people themselves, talking, thinking, coming to personal decisions on the problems of their community – are the material for an exciting and entertaining television programme.'

Unfortunately such programmes are still exceptions. It had been hoped in England, France and the US that cable television would play a part in the democratic exchange of ideas, but these hopes have so far been disappointed. In England five experimental local television stations were very successful and demonstrably contributed to strengthening public and community spirit; but the experiment had to be discontinued for lack of funds. In New York City the critical television institute of Columbia University, Network Project, published the following report on the present state of affairs:

> The few community video centers which attempted to serve neighborhoods were doomed by lack of financial support. . . . Cable companies are equipped to meet minimal demands for access but not to create or encourage effective use of video. Nor have they encouraged the growth of audiences for public channels. For New York City's population of eight million, 'Sterling Cable' offers two partapaks, no studio, no advertising in newspapers, on television or billboards and no program information. 'Teleprompter' has a studio with one cameraman open on weekdays but no portable equipment. . . . Neither company permits live studio transmissions.

The analysis made by the New York institute tries to find reasons for the disappointing results and comes to the following conclusion:

> Yet the story of cable goes far beyond this. Like other media in our society, it has fallen victim to a fundamental misunderstanding concerning technology. Technology does not exist in a vacuum

and is not an end in itself as media 'freaks' and entrepreneurs would lead us to believe. It is surrounded by a complex economic structure which has as its primary goal the maximization of profits.

It is to be feared that the technical possibilities of video and cable TV will first of all strengthen and improve the 'targeting' of advertising and propaganda aimed at potential consumers. Observers with an interest in democratization are therefore inclined to regard the further development of electronic media as another step towards technocracy. This aspect must be borne in mind when considering the other ideas and experiments that I shall describe. They could and should be the start of more participation by Everyman, but they also open up fresh possibilities of manipulation from above.

Contradicting the oracle

Would it not be possible as well as desirable to discuss problems of national and international significance with millions of people? Ideas such as this were first voiced by the American inventor and television chief V. Zworykin. He thought it would be possible to add an 'opinion control' to every radio and television set: different opinions could be telephoned through to the relevant stations both during and after a programme. The idea was taken up by the American systems analyst West Churchman: together with the Germans Helmut Krauch and Horst Rittel he prepared a programme on American research priorities sent out by the radio station at Berkeley, California.

The problem was formulated as follows: should we go on pouring billions of dollars into armaments research, or would it be better to spend the money on improving education and living standards in the developing countries or on trying to solve other non-military problems? The programme differed from others in that it had a data bank with facts and figures about the latest developments in armaments, the number of engineers available among the big power states, the gross social product in various countries, their military expenditure, and much more besides.

Listeners who had applied to join the programme were sent an objective written account of the question to be debated, together with extensive statistical information and even political cartoons to clarify the conflict. They could telephone during the transmission and say which arguments they supported. Their comments were quickly evaluated by a computer and transmitted to listeners in the course of the debate. They were also able to telephone in questions and opinions after the programme.

Krauch is the founder of the Study Group for Systems Research at Heidelberg, one of the first institutions for solving problems in sociology. From this beginning he developed the ORAKEL (oracle) system, which paradoxically aims to ban self-important announcements by the ruling political establishment and to replace them by discussions among wide sections of the public. Krauch thinks that some sections of the population are badly represented or not represented at all in parliament, while others are over-represented. In the television debate, on the contrary, representatives from every social group directly or indirectly affected can take part provided they know enough about the subject.

But it seems to me that here the anti-democratic 'qualification barrier' comes down and excludes a large number of the people concerned. Krauch calls his debates 'organized conflict' because, after conscientious preparation, the opposing points of view are to be sharply and clearly presented. The viewers can telephone in their opinions throughout the programme. They have to give their sex, age, profession and educational level, and vote in one of five categories:

5 = strong agreement
4 = agreement
3 = neutrality
2 = opposition
1 = strong opposition

In 1969, following three radio programmes on the subjects of long hair among the young, female emancipation and data banks, West German Broadcasting transmitted its first 2½-hour television 'Oracle'. In order not to leave the composition of the telephone participants entirely to chance, a representative panel of twenty-five citizens was set up. The programme began with a cartoon film illustrating the German parliamentary system. Then there was a twenty-minute film on the theme of the discussion, environmental pollution.

Not until then were the debaters as well as the viewers asked three questions, to which supplementary questions could be added later as the discussion developed. The three questions were:

1 To what extent is the pollution of the environment the inevitable price of progress?

2 To what extent is industry prepared to adopt voluntary measures to safeguard the environment?

3 How many people would be prepared to pay an extra 10 per cent on their taxes to solve the problem of environmental pollution?

An essential part of the programme was the data bank which acted as an adjudicator. It was able to confirm statements, correct errors and answer factual questions.

Krauch's report on the experiment shows his commitment:

Would anyone telephone at all? The television people were obviously anxious and hoped that at least a few loyal friends would telephone. The representative from industry thought the second question unfair and said that even the third question obviously sought to put the whole blame on industry. But then we began to hear – relayed from the computer centre – hundreds of telephones ringing. And then it was time for the organized conflict to start. As we had expected, air pollution was the first topic. The representative from industry accused the consumers of contributing more than half of all pollution with their car exhausts and domestic fuel. The data bank did not contradict him. The government representative gaily reported several counter-measures already taken by his ministries, the left-wing poet drew attention to the high profit rate in industry, the systems analyst tried to point out the connection between consumption and advertising. The professor of medicine representing the government health service found the industrialist personally very agreeable and largely backed his views. This brought him the first rebuke from the panel: would he please remember the interests of society as a whole and not just of industry, because after all, his salary was paid by the government . . .

The thirty telephones began to be busy while the 'conflict' was still going on and they went on ringing until well after midnight. Some viewers who lived near the Cologne studio and found the telephone engaged were so keen that they drove to the studio in cars or taxis and tried to get in. Others rang the private numbers of the producer and director.

The debate continued on the following evening. Krauch remembers: 'This time over a thousand viewers telephoned. Seven hundred and forty-one had seen the previous day's programme. But even among the second group a large number wanted to take part in the next programme. They had discussed "Oracle" with their friends and relations and were therefore able to form an opinion. Altogether nearly three thousand calls were received.'

Containing the influence of computers

A special number of the New York magazine *Radical Software* was devoted to the subject of the democratic possibilities of cable television. One note of warning was sounded among the many enthusiastic voices:

> And as always, every communications device finds military and police applications. If telephone tapping is a fear, imagine the surveillance made possible by coaxial cable. Even without tapping, the telephone company now has a complete record of every long-distance phone call you have ever made from your home. When virtually *all* information an individual receives is processed through a cable, privacy will become a nostalgic memory. By pushing a retrieval button, controllers will know what movies you have selected to see, what books and magazines were printed out for you, at what part of the news you lost attention, who you spoke to and what your facial expression revealed about your attitudes . . .

This pessimistic view of the future of the media is exaggerated, but it contains a justifiable anxiety which can apply to any technical invention: it can easily be manipulated and used for purposes quite other than those its inventors intended.

I have described a few attempts to use information techniques for revitalizing and improving democratic procedures. But I cannot share the vision of millennial man as he appears to many electronics enthusiasts: crouching for hours, more or less alone, by his television screen and speaker and helping to govern. It would be bad for the future of human freedom if most or conceivably all information came across mechanical systems, and those very qualities which we regard as human would atrophy: spontaneity, imagination, warmth, proximity, intimacy.

But some people believe that everything can be done by mechanical means. Their idea of 'total information' is typified by the Japanese cyberneticist 'Donald Kenzotaki' and his man–machine system 'Intersex'. His report appeared in the British journal *Architectural Design* in September 1969 and contains the following passage:

> The object of the experiments is to gather data on a possible system for the remote communication of the complete sexual experience, through feedback communications technology. Each

partner is able to see the other on the various closed circuit cameras, feel the sensations the other is undergoing through the system of remote sensors, exchange verbal data via the telephone circuits, and to control the other partner's manipulator through telechiric transmission of pelvic and other body movement. Through a system of PCM (pulse-coded modulation) the four visual channels, the telechiric instructions and the sensory information can be transmitted in both directions simultaneously on two 6-megaherz channels. We are further hoping to reduce this to a single 6-megaherz channel, so that we may use standard microwave relays and eventually satellite transmission, so that partners, however they may be separated physically, may continue an intimate relationship.

He goes on to describe how tactile experiences can be stored on magnetic tapes or 'tactile plates' and lead to 'relations' with electronically recorded sex stars. But we do not need to follow the inventor of long-distance copulation further in his elaborations: it is obvious that he is deliberately being absurd in order to lay bare the mechanistic, anti-humanist root of certain kinds of cybernetic developments.

His account produces a mixture of fascination and horror, and that was presumably what the author intended – although he was not really a Japanese at all, but most probably the banker Oliver Wells. Wells's hobby – which soon turned into his main occupation – was speculation on cybernetics, its philosophical basis, its consequences and its probable further development. He could hardly have found a better way to expose the transgressions of information technology than through this 'love story' told in cold technical terms. It first appeared in a publication, addressed only to a small esoteric circle, called *Artorga* (artificial organism), without any commentary at all. More widely circulated publications picked it up and apparently took it literally without bothering to think about the contents.

Like love, democracy needs direct encounters whose message must not be watered down or adulterated by the media. Cameras, the electronic spectrum, satellites, cables, magnetic cores, data banks, receiving apparatus of every kind are only vessels and transport for information and must not be allowed to determine the contents. The free creation of social concepts, the formulation of plans, open discussions as to whether and how they are to be adopted, can never quite succeed in conversations between two-dimensional shadows. The electronic market-place and the global village can only be humanistic concepts so long as they do not try to take the place of the real market-place, of real communities or of real contact between people. They can only act as auxiliaries.

But it would be both possible and desirable to extend the knowledge and judgment of every individual and every group by means of information storage, information comparison and the combination of different kinds of information, providing them with a better chance of coping with increasingly complex conditions:

◇   J. R. Schade suggests that there should be a 'health co-ordinator' in every neighbourhood, on the same system as the fire alarm. It would be a transmitter/receiver linked to a data bank: in cases of sudden illness it would provide first-aid advice and details about the nearest chemist, doctor, and hospital with available beds. A more highly developed version would enable the public to ring the information bank from any telephone and to receive answers individually adapted to individual questions. Similar systems could provide legal and consumer information, and other forms of assistance.

◇   Nicholas Negroponte of MIT has built what he calls an 'architecture machine'. It is a computer fed with facts and suggestions for the non-expert who can discuss his living requirements with it; it can correct his amateur drawings, cost them, and even point out the advantages and disadvantages of his ideas and designs.

◇   Professor J. S. Saloma is of the opinion that democracies should have open, largely computerized administrations, which would give every representative direct access to all the information of the executive and its organizations. The concept might even be extended to include every citizen: in the end this would deprive the paper-hoarding bureaucracy of its power.

◇   The anthropologist and cyberneticist Maguroh Maruyama of San Francisco State College goes even further: he thinks that radio and computers could help us to adopt non-hierarchical, non-authoritarian decision-making procedures such as the Eskimos and Navajo Indians use without ever failing to reach unanimous decisions.

However, up till now the development of information technology has tended in the opposite direction, towards greater concentrations of information and power. The technical means determine the ends. An example of this was the strategy in the Vietnam war: politicians and soldiers relied so much on their data banks that they lost sight of reality and 'abdicated their decision-making responsibility to a technology they don't understand'.

This quotation comes from a lecture by Professor Joseph

Weizenbaum, who, at a meeting in Vienna, described how the generals acted according to the computers and 'lost' the original intentions of the Pentagon for bombing 'especially worthwhile objectives'. Weizenbaum declared:

> These often gigantic systems are put together by teams of programmers, often working over a time span of many years. But by the time the systems come into use, most of the original programmers have left or turned their attention to other pursuits. It is precisely when gigantic systems begin to be used that their inner workings can no longer be understood by any single person or by a small team of individuals. This situation has two consequences: first that decisions are made on the basis of rules and criteria no one knows explicitly, and second that the system of rules and criteria becomes immune to change. This is so because in the absence of detailed knowledge of the inner workings of the system, any substantial modification is very likely to render the system altogether inoperable. . . . No human is any longer responsible for 'what the machine says'.

A similar warning is sounded by Rex Malik, a critical English publicist specializing in social problems created by the advance of the computer. He suggests that all printouts from computers should be accompanied by the following warning: 'Not all the experts in the field agree with the data on which these conclusions are based.'

In this connection I must mention Harvey Matusow, a jazz musician with scientific training who emigrated to England from the US for political reasons and founded the Society against the Misuse of Computers. At first he concentrated on disseminating suggestions for sabotaging data machines by technical means. When this became too difficult, he invented other methods which, he assured me, were harmless but effective. He holds that a great deal of the communications, requests and questionnaires that every citizen receives are already the result of computer calculations. When he is approached by people to fill out questionnaires or other similar documents, he sends them a questionnaire of his own and asks them to let him know who they really are, how they happened to pick him out, and what use they plan to make of the information they have asked for. In most cases he hears no more of the matter.

This is an original method of containing the influence of the computer, but hardly one which is likely to be successful in the long run. Just as there are sworn auditors, there may have to be sworn computer programme controllers. But in the long run it will certainly be necessary to balance the computer networks of government and industry by citizens' networks whose data will be collected and grouped according to other premises, other interests and other goals. Data banks for parliament, for instance, will have to be fed with – among other things – data that will not be found in the data storage systems of the ministries. Every group with social interests will have to make use of this new machinery if it is to be up to date with discussions and not to be swept aside for being insufficiently well informed.

Critical journalists writing for newspapers, radio, television or topical paperbacks will soon be joined by critical reporters whose medium is the data bank. For a long time now the media, with their chronic shortage of time and space, have been unable to find room for the wealth of facts that need to be considered before any political decisions can be made. We need independent or opposition information banks kept up to date by correspondents; only with their aid can the citizen fulfil his duty of criticizing and controlling events in an 'information society'.

Centres for the community

For some time it has been possible to arrange combined television/data bank discussions over distances of hundreds and thousands of miles. Fifteen American computer research centres belonging to the Pentagon's ARPA (Advanced Research Project Agency) work together as closely as if they were all gathered around the same green table for their daily meeting. But in fact some are on the West Coast watching a big screen at the Stanford Research Institute, others are on the East Coast at the Lincoln Laboratory near Boston, and a third team is in the Midwest at the University of Illinois.

What distinguishes these conversations from the usual telephone discussions is chiefly the fact that the data banks are all linked together so that their data complement and strengthen each other. Within seconds they can retrieve facts required for the discussions from storage and project them on to the screen. Instead of being confined to the papers and plans they happen to have brought in their briefcases, members of a discussion group can press a button to have the whole of their archives at their disposal. J. R. Licklider, who developed the system together with Doug Engelbard and Robert W. Taylor, thinks it will soon be possible for scientific teams to co-operate closely and constantly across the whole planet with everyone having access to the same pooled information.

An 'on line community' of this kind would be miles apart physically but closely linked by its interest in the same subject. Its members would be much more closely in contact than they could be by telephone, telegrams or teleprinter. Nuclear research laboratories, observatories, social scientists and the university institutes of many countries could unite on common projects with much greater ease and frequency.

But amazing prospects are opening up for contact among ordinary citizens as well. The imaginative architect Yona Friedman of Paris thinks that before the turn of the century it will be possible to hold world congresses without any of the participants leaving their bases: the meetings would take place in the respective communications rooms. Peter C. Goldmark, for many years head of the experimental section of Columbia Broadcasting, has worked out a plan for connecting the inhabitants of Connecticut electronically with their places of work, so that they would no longer need to travel to and from their offices in the rush-hour traffic. Even an electronic town hall has been planned by a group from the Center for Policy Research under the leadership of Amitai Etzioni and Eugene Leonard. Their 'participatory technology system' is called MINERVA and envisages constant discussions between geographically separated citizens over a special radio and television network.

But electronic aids cannot blind us to the fact that the citizen of today has fewer and fewer places where he can meet his fellow-citizens informally and in person. The market-place, the piazza, the corso have not been replaced by any new type of concourse. If there is to be more democracy, there will have to be more opportunities for people to meet.

Pedestrian precincts, play streets and recreation centres are already being planned in more and more towns. Where they already exist – in Rotterdam, Dronten, Munich and Frankfurt, for instance – they have tended to attract chiefly strollers, street vendors and in some cases (as in Santa Monica, California) also hoodlums: they have not yet become agoras in the Athenian sense where public matters are discussed. What these free spaces lack is the higher purpose which in the past radiated from the churches and the town halls.

The community spirit is reawakening; it can be seen in the wave of community projects, street demonstrations and improvised discussions; if it becomes sufficiently strong, it should also find an expression in communal buildings and institutions. Today Everyman is only tolerated in most civic and state institutions, hemmed in by rules and regulations and made to behave properly by stern guardians of public property: in the new institutions he will feel free as the master in his own house.

In the next few years we can expect to see a large variety of projects in which the human spirit will be set free, human creativity will reawaken to opportunities for activity and recognition, and man's right and ability to decide his own fate will be acknowledged. Among them are:

◇   public audiovisual libraries such as the one that already exists in Stockholm and the one opening soon in Frankfurt;

◇   'community information expositions', as pioneered in 1972 by T. V. Vonier and R. A. Scribner on the occasion of the annual conference of the American Association for the Advancement of Science (AAAS); the experiment was called 'Capital City Readout', and the aim was to set up similar exhibitions in museums, parks, pedestrian precincts, empty shops, etc., where social problems could be plastically presented and illuminated by means of planning games;

◇   *maisons d'art* (art houses) like the one in Caen, or open museums like the exemplary Louisiana-Museet in Copenhagen, the Museum of Modern Art in New York, and the Museum of the Twentieth Century in Vienna, where the viewer is drawn more and more into becoming a 'doer';

◇   permanent rent-free performing places and theatres for theatrical troupes, bands, amateur actors;

◇   discussion corners in busy streets;

◇   information centres where critical independent observers will present current affairs in words and pictures, possibly supported by data banks owned by the people themselves – these presentations to be different from those put out by official channels;

◇   popular festivals, to be held once or several times a year, like the Old Town Festival in Hanover, Germany, which attempts to bring people together in a light-hearted way under the motto 'joining in, playing, experiencing, touching, eating, drinking, feeling, hearing, smelling, seeing, tasting'.

To start with, the new areas of public activity will depend on 'animators' who will awaken the citizens from the passivity to which they have become accustomed and will teach them how to express themselves and how to take action on their own initiative. If they have as much verve, charm, humour and psychological insight as the *gentils animateurs* of the 'Club Méditerranée', then they will soon succeed – as experience has shown – in activating the timid, the frozen and the silent.

Today's communities have no centres for radiating spiritual and emotional values. The role could be taken by 'houses of the future', where the citizens could discuss together their hopes and fears, their ideas, plans and suggestions. There would be exhibitions where amateurs, artists and specialists could show their prognoses and visions of the world of tomorrow. Simulations of future conditions could be enacted with people playing different parts, and discussions would follow. These institutions should receive a constant flow of new ideas and inventions which could then be presented, judged, examined for their probable consequences and publicly discussed. There would also be workrooms for 'planning cells' in which citizens could come to terms with long-term problems: this idea was first suggested by Peter C. Dienel of the *Gesamthochschule** at Wuppertal. The aim of this new 'social invention' would be to decrease the information gap between the decision-makers and those affected by their decisions. Citizens for 'planning duty' would be chosen by a random process, as jurors are. They would spend several weeks on a crash course preparing them to participate in planning, and would then take part in the actual process. Any loss of income they suffered would naturally have to be made up. In the 'houses of the future', children and students would not only be admitted, but would receive privileged treatment. No proposal would be turned down in the first instance as impossible, or branded as crazy, or deemed unreasonable by false ideas of what is reasonable. The as yet 'unacceptable' would here be rewarded; the rule would be courage, not conformity; not harmony with existing conditions and values, but a challenge to them in the spirit of the future.

I attended a conference on the future at Honolulu. It was called 'Hawaii 2000' and it taught me how stimulating it can be for the citizen to invent and imagine the future. The conference had been preceded by six months of preparation: six hundred citizens of every race and class had been divided into ten working groups and had discussed the politics, economics and civilization of their island state. The wish was frequently expressed that this temporary forum of civic discussion should become a permanent feature. The fact that this expressed hope was not heard by the State administration only shows the existence of that fear of liberating the social imagination which we have already described.

When we consider how to renew, widen and revivify democracy, we must not be content with spontaneous community projects, with television guerrillas, occasional 'Oracle' transmissions and town-planning forums; we need permanent institutions where the

---

* A new type of comprehensive university.

general picture can be grasped, discussed and enriched with further ideas by all those who are interested. Otherwise all these experiments in democracy will remain outside the mainstream, sunk in provincialism and superficiality. They will be no more than fashionable political spectacles, and their chief purpose will be lost – which is to bring as many thoughtful and creative people as possible into the continuing process of making society more humane.

The first experiences of trying to increase opportunities for many or even all citizens to participate have often been difficult. They have shown that, almost two hundred years after the revolutions that were meant to put power in the hands of the 'sovereign' people, the vast majority – even in the Western democracies – have not even learnt to exercise their democratic rights. Citizens have continued to be led or misled; leaders have seized control; sometimes the diffident citizens have thrust it upon them; criticism and discussion have been thwarted in the name of a false unity which made it impossible for people to act as individuals and to make their influence felt. Real democracy in the sense of participation by Everyman is only just beginning. I have briefly described a few clumsy and unproven experiments which are the first steps towards fulfilling the promise that has been awaiting its fulfilment since the eighteenth century. Today the greatest obstacles are the economic power groups who try to subordinate the many to the interests of the few – often amid hypocritical invocations of the right to participate. In the future they will have to be prepared for increasing pressure upon their positions.

One of the chief barriers in the way of making society more democratic and humane is high technology: it enables the mighty to centralize power, manipulate people and pollute the environment, and it threatens to bring about catastrophic wars. Man at the millennium must give first priority to controlling and changing this dangerous apparatus, which has the evident potentiality for promoting non-democratic or even anti-democratic rule.

5 Open Man

The great reversal of values

Most representations of the man of the future show him with in-human features. His face is hidden by a mask such as is worn by firemen, welders, policemen, deep-sea divers and astronauts. The mobile living body disappears beneath shapeless armour which differs from medieval coats of mail only in its material and in having cable attachments and antennae. The living creature is carefully locked up inside and has to carry its prison around with it. The armour can only be exchanged for painfully tight containers like tanks, submarines or space capsules. For only thus – protected and imprisoned at the same time – can the man of the future survive in a hostile environment.

But this picture tells us more about the recent past and a waning present than about the future. It is the portrait of the aggressive and defensive *Homo technologicus* of the mid-century. He will become more and more an outsider, the relic of an age given to violent ag-gression and conquest, the Don Quixote of a new era which will have little use for this kind of heroism. Presumably millennial man will not look very different from today's man in the street. He may have gayer and more imaginative clothes. But in spite of his greater interest in public duties he will not deny his individuality by wearing any kind of uniform: on the contrary, he will stress it by his unin-hibited and natural behaviour. Masks of every kind will be seen through or out of fashion.

In spite of occasional relapses, a long-term trend towards the re-versal of accepted values and aims has been evident throughout the world:

◊ from isolation to openness;
◊ from conquest to expansion;
◊ from producing to experiencing;
◊ from forced achievement to free development;
◊ from hard to soft;

◇   from rigid to flexible;
◇   from useful to playful;
◇   from death-dealing to life-giving.

New forms of working together are being tried out, where the
partner is regarded not as a rival but as a support. Fewer people now
strive after more power and speed, more after greater sensitivity and
deeper understanding. Longings which only yesterday (and perhaps
even today) were despised as the ridiculous desires of petty bourgeois
aesthetes, like the longing for peace and beauty, have become a
universal desire. The psychologist Richard E. Farson, a former
director of the California School of the Arts, demands new constitutional rights for all. Among them he lists the rights to leisure, to truth,
to sexual fulfilment, to peace, to one's own individuality. Nicholas
Rescher and Kurt Baier of Pittsburgh University try to represent the
noticeable changes they have established in the course of their work
in 'value research' by means of a list of possible 'value changes' in
America; the following examples strike me as particularly interesting.

As a response to population increase and urban crowding they
anticipate in the first place:
1 devaluation of privacy;
2 strengthening of small-group values to provide foci of 'identity'.

As a result of improvements in means of traffic and communication:
1 upgrading of mankind-oriented values (and to some extent a
   corresponding devaluation of nation-oriented values);
2 Growth of cosmopolitanism and strengthening of internationalism.

As an apparent reaction to the 'Big Brother' state:
1 probable upgrading of democratic values (in face of obvious
   threats thereto);
2 ambivalence to authority.

As a possible consequence of the increasing sophistication of
modern mass-destruction weapons:
1 downgrading of the characteristic species of national pride;
2 upgrading of mankind-oriented values.

Humanistically oriented values, which were reasserted at the end of
the 1960s, also dominate by far the later investigations of Daniel
Yankelovich, although this leading American social scientist notices
a 'turnabout' in many other fields which in general points to a
retreat of political radicalization. Nevertheless, the following stand
at the head of his list of 'welcome social changes':

1 more emphasis on self-expression (76 per cent of non-college and 83 per cent of college respondents);
2 less emphasis on money (74 and 80 per cent).

As for his findings on 'very important personal values', the leaders are:
1 love (88 per cent non-college, 87 per cent college);
2 friendship (87 and 86 per cent).

The strong emphasis on the right of self-expression, as found by Yankelovich, is also confirmed, after the decline of the 'youth revolt', among non-academics as much as among students. It is, however, the view of young American social scientists (such as the group around the political economist H. Gintis at Harvard) that the spiritual and cultural development of the individual can only be accomplished within the solidarity of a community that has renounced egotistical urges such as careerism, the pursuit of profit, and consumer greed. The true higher and further development of the individual proceeds slowly and imperceptibly; it requires sacrifices of money and time which a materially oriented society cannot often afford to make. Besides, régimes that are primarily concerned with increasing production, be they Eastern or Western, have no real use for developing man, whatever they may declare to the contrary; for as he develops, so he becomes less politically dependable. In a commentary on the occupation of Czechoslovakia, the late Ernst Fischer depicted the powers-that-be regarding developing man as 'too dangerous an invention: he suffers from imagination'. In spite of his despair at the brutal suppression of the Prague experiment, Fischer expressed this hope: 'But it is not so easy to abolish man. . . . He is beginning to rebel against his dehumanization.'

The rebellion has many facets. If it confines itself to negating the values it abhors, it can soon develop into their mirror image. Advances in new directions are needed. And this is where I would stress the significance of various attempts to counter the dangers to which our species is exposed with a new, comprehensive 'science of man'.

The beginnings of a 'science of man'

Seen from the Pacific, the Salk Institute for Biological Studies at La Jolla, California, with its grey concrete walls, looks like a commando bunker on a clifftop. On closer view, attempts are revealed to counteract the heroic architectural style. A windswept piazza opens out. It was intended as a meeting-place for the research workers.

On the day of my visit it was empty, and I was later told that this was usually the case. The same emptiness reigned in the long passages and behind the huge windows through which the laboratory instruments are visible. Not a soul even in the library.

The founder, director and inspirer of this remarkable research institute is Jonas Salk, the discoverer of the anti-polio serum. He is a Nobel prizewinner, philanthropist and philosopher. He believes in the 'biological revolution' which, at the critical moment in history, will produce the necessary life-oriented change in mankind. A leading tenet of his life's work is that man, in many ways, is 'incomplete' – in self-awareness as well as in his development. There are still areas within man and in man's relationships that remain to be investigated.

A glance at the annual reports of the Salk Institute corrects one's first impression of defensive isolation. Within the last decade numerous ideas and initiatives have gone out into the world from this centre. Salk knows how to attract not only biologists and geneticists, but outstanding personalities from the most varied intellectual fields and disciplines. Leo Szilard began his last work at La Jolla; Herbert Marcuse lives nearby in a modest house on the beach, and comes up to join in discussions; Jacques Monod worked here on his successful and controversial book *Chance and Necessity* (New York, 1971; London, 1972); Edgar Morin of Paris, in his search for new foundations and extensions for sociology, was able to gain fresh experiences in what he calls 'this Sinai blockhouse': biology, anthropology and cybernetics enlarged his point of view. Here, at the southern tip of California, the foundation is being laid for a 'science of man' uniting all the research disciplines concerned with the human condition.

After various intermediate stages, these beginnings have already led to the setting up of the international 'Centre pour une science de l'homme', founded in November 1972 at the old abbey of Royaumont near Paris. Its function in the history of intellectual development could become as stimulating and pivotal as that of the Cavendish Laboratory in Cambridge, where Rutherford laid the foundation of modern atomic science.

The position of *Homo sapiens* on the threshold of the millennium is becoming more and more critical, and it is therefore fairly safe to predict that large numbers of institutes, seminars, project groups and experimental undertakings will spring up for the intensive study of man and the possible improvement of his chances. Genetic experiments dealing with man as a conglomeration of cells will take second or third place to the examination and development of those faculties which distinguish him from all other beings. Morin, for instance, is convinced that peaks of imaginative creativity such as can be seen at their earliest in cave paintings are natural and peculiar to man. Man's attitude to time is also unique: he is able to think beyond his own life-span, as can be seen from the earliest burial cults. And his special hazard, irrationality, can save him under certain circumstances, for it is a protection against too much rigidity.

I imagine that the new 'open system' man will go on being tested and developed, especially in 'mind laboratories', which will become as important in the twenty-first century as were the physical and chemical laboratories in the preceding period. In these new research stations the distinction between the disciplines will be less and less significant. The spirit of rank-pulling and dogmatism so typical of many of today's scientific establishments and gatherings will not be able to maintain itself in the new era of co-operation. At congresses people will no longer read papers completed and sent in months beforehand, which are then politely pulled to pieces; they will bring with them new, unfinished ideas which can be developed in collaboration with other minds. Artists will presumably combine forces with scientists to carry forward the advance into unknown territories. Inter-relationships will be too complex to be understood or solved within one narrow branch of culture: we may see a type of interdisciplinary pioneer developing, a specialist in several fields who will develop hitherto unknown powers of synergy and synthesis.

Information specialists and cyberneticists will also play an important role, the former because of their ability to hold the ever-growing number of data at their disposal, the latter because of their experiments in grouping data in various combinations, thereby creating new patterns of hitherto undiscovered interaction. Even new personality models might be drawn up in this way. Such simulations on the system 'man' could help the computer to discover new blank areas on the map of the inner universe and thereby point to useful directions for the expansion of human potential.

The science of man can be misused and abused; one can imagine cases of violation and destruction as perilous as – or perhaps even more perilous than – physical threats to life. All those engaged in this tricky work are well aware of the dangers. But they feel that they have learnt from the experiences of the atomic scientists and that they are more prepared for the temptations and dangers ahead than the physicists ever were.

There is a place here for doubt, and it was probably this doubt that made me find the external aspects of Dr Salk's institute so sinister, in spite of all the great ideas and aims that emanate from it.

With a few exceptions the avant-garde – not only in science but in art and literature too – is so isolated that its isolation must be a peril. The democratization of culture means not only the awakening of the blocked potential of many individuals, but also a defence against the one-sided exploitation of knowledge in the service of a small élite. If the promising and potentially fruitful attempts to develop man are to lead to greater humanity rather than to genetic engineering, then the pioneers in the field must avoid the mistakes made by other scientists and 'go to the people'. Many intellectuals still regard this step as a waste of time, a submission to the second-rate. It is just such deep-rooted prejudiced attitudes that make it essential for the science of man to explore the possibilities of collaboration between persons from different sections of society. Intellectual pride, however supposedly reasonable and sustained by so-called facts, is the greatest obstacle to further development in a direction beneficial to mankind.

The new openness

'Fundamentally, we are all lonely.' This was the view not of an old person, but of a young, good-looking woman. In answer to questions she said she was a doctor and had emigrated years ago from Czechoslovakia to Munich. She was taking part in an open discussion group which met every day during the 1972 Olympic Games. Everyone agreed with her, although the participants in the discussion had met quite by chance. Passers-by on their way to or from the various events would stop out of curiosity at this group assembled at the foot of a flagpole. They would listen and then usually pass on. But if they happened to be interested in the subject under discussion, they would sit down on the grass and join in.

The real purpose of these meetings was to discuss the world of tomorrow. Most of the discussions took place after the provocative performances given by the Berlin mixed media group 'Olympia – Richtung 2000' (Olympia – Towards 2000); they were meant to be a commentary on the terrifying visions that had been enacted. But contrary to what had been intended, the theme of present isolation and alienation often tended to push aside the great subject of the future. I would ask these people from all classes and all countries what was their dearest wish: not one mentioned a technical invention. Frequently they wanted more interesting and meaningful work, oftener still a fair share-out of property and responsibility; and most frequently of all they wanted more togetherness, more personal proximity, more human warmth.

I think that the strength of this desire and the depth of misery from which it springs have not been taken seriously enough by social diagnosticians. The cities of today are not merely 'inhospitable', as the social psychologist Alexander Mitscherlich said. The word is much too mild; even in negation it sounds quite cosy. The style of our 'silos for living', our 'prisons with television and telephone', our 'parking lots for people' is indicative of a serious social malady. There are innumerable diagnoses for it, but so far only a few successful cures.

Once the cry was: 'City air sets you free!' Now it should be 'city air makes you lonely', 'city air makes you sad', 'city air makes you sick'. That is why we see thousands of communal experiments, beginning with packaged tours, whose alleged object is economy, and leading through group therapy and sensitivity training to schools for parents, community projects, communes and extended families – many of them helpless attempts to restore human contact. But most of the participants have to return to their solitary confinement, to their professional roles, to their so-called normal lives which, after only a week's successful 'human potential training' in a group, can seem as insane as they really are. Jane Howard, a reporter for *Life* magazine, describes her return from group life to the office as follows:

> At the Boston airport the reservations clerks seemed impersonal as robots, the skycaps avaricious and the newsstand lady grouchy. My fellow transients looked harried, with hunched shoulders and furrowed brows. I'd have looked that way too had I not just emerged from eight days of exquisitely heightened sensitivity. As it was, I was absurdly aware of the humanity of all these strangers. I had to restrain myself from gazing with what would have seemed awful intensity into their eyes, and administering unsolicited backrubs to the ones who looked tired and tight. If I had yielded to these impulses I would likely have been carted away to some institution, as they say a girl once was from the San Jose bus station after a workshop at Esalen.

'Group dynamics' was started in the US towards the end of the Second World War by a German refugee, Gestalt psychologist Kurt Lewin. Industry took it up in a big way, and 'progressives' have seen this as a sign that it is not a step towards greater humanity, but 'part of a constantly developing technology to increase human achievement and human value from the capitalist point of view of higher profits'. The answer to this objection is that if such methods lead to higher achievement in factories and offices, then that is good for the employee as well as the employer. But not only because he may be better paid: also because he will have achieved more contact with his colleagues. Before the introduction of group dynamics he may have known them only superficially. But encounter groups force the participants to discard their armour, and project teams teach people to work together at a task: the result is the development of the conditions necessary for solidarity in action. At first the new intimacy was often ridiculed because people felt embarrassed; but in the long run it often turned out to be more deeply and firmly anchored than the former community of material interests, whose foundations were frequently somewhat shaky.

During the last few years large numbers of new groups with new functions have been formed all over the world. They exist alongside the old groups such as religious communities, professional bodies, clubs, political parties and labour unions. I have not been able to find any survey of them. But I think we are faced with an important phenomenon in man's striving for further development, and so I have undertaken to list the new groups under various headings:

I Predominantly practical motivation
a) Tenant and housing associations, neighbourhood groups
b) Consumer groups
c) Car pools
d) Group holidays
e) Kindergartens

II Predominantly spiritual motivation
a) Therapeutic groups
b) Group marriages
c) Encounter and sensitivity groups
d) Meditation groups
e) 'Island communities' turned inwards for mutual protection

III Political and idealistic motivation
a) Community projects
b) Action groups
c) Socialist communes as examples and bridgeheads
d) Children's co-operatives

e) Agricultural co-operatives
f) Ad hoc groups for solidarity

IV Predominantly professional motivation
a) Planning offices
b) Media production groups
c) Doctors' group practices
d) Pensioners' work groups

All these are attempts to fill gaps in social systems that have grown too large and impersonal. But beyond that, all these associations with their totally diverse *raisons d'être* have one common characteristic, an attitude typical of our day which has come to be taken for granted with amazing speed: I mean the great degree of openness between individual members. Secrets which people hardly dared admit to themselves, which in the past would have been confided only to a member of the family, a close friend, priest or doctor, now burst forth in a circle of people who frequently have only just met.

The new openness is not confined to therapeutic gatherings, encounter and sensitivity groups: to a greater or lesser degree it occurs in other groups as well. People probably overestimate the degree of sexual promiscuity involved, and underestimate the degree of psychological intimacy.

This phenomenon is of the greatest significance for a prognosis for humanity. For a great change will occur if ruthless, painful and self-liberating openness becomes the rule and people stop playing roles, wearing masks, hiding their guilt and their desire, practising daily self-censorship, living with lifelong shame and silently submitting to mutual deception and apparently ineradicable hypocrisy.

Objections may be made; scepticism is not out of place. A change in such deeply rooted instincts and habits cannot come suddenly, without backslidings or mistakes. Above all it must not be imposed from above or from outside. That would only lead to new tactics of concealment and flight which psychologists from the University of Michigan have called 'Promethean behaviour': a reaction – typical for our times – against the intrusion of public authorities into private spheres.

But so far no such pressure exists. Opening up, confessing, exposing social lies, tearing down repressive obstacles – all these activities which used to take years of conversation with the psychiatrist – are now practised with brusque determination in groups. Not only the specialist may feel uneasy. Perhaps this psychological striptease will do more harm than good? Perhaps it is simply a question of learning new disguises, new forms of hypocrisy, and false sincerity?

These doubts are justified. But they do not alter the fact that these experiments are gradually changing the climate of life, so that acts of self-revelation are considered less and less exceptional, shocking, and therefore dangerous. And man's gradual opening up is in the line of historical development. Once it was impossible to leave one's house unarmed. Nowadays it is normal in most civilized countries. And in spite of relapses, there are fewer walls, fences and frontiers. Society has invented security measures which enable people to live more freely and with less anxiety than in former centuries.

It is therefore decidedly possible that 'open man' will have his chance in the next decades. In spite of many trends in the opposite direction, he may be able to demolish repression of every kind. But if that is to happen, then more will be required than mere spontaneity. There will have to be years of slow, careful work with an increasing number and variety of groups. Moreover the behaviour of large and small groups will have to be studied by ethnologists, social psychologists, anthropologists, sociologists and peace researchers. We are still at the beginning of open human co-existence without the constant fear of being overpowered or overtaken. There has not been nearly enough research in the field.

As an example of practical group training in 'openness' and of a 'fraternal society' I should like to mention the work of Richard Hauser and his pianist wife Hephzibah, the sister of Yehudi Menuhin. They work chiefly with the socially deprived and try to establish with them an open solidarity in which everything can be discussed. It is interesting to summarize the biography of this typically millennial man.

He was born in Vienna in 1911 and studied sociology and psychology. In 1938 he emigrated to England. He founded and ran a school of social studies in Rome. During the fifties he worked with the handicapped and with immigrants in Australia, where most of the refugees from Hitler were taken during the Second World War. In England he worked with prisoners and prison personnel, in special schools and with West Indian minority groups. He advised Black leaders and acted as a mediator. In 1962 he began working for the homeless in West Germany.

What is so attractive about Hauser's writings on his group experiences is the fact that they are addressed not to specialists, but to 'people who like trying something new, people who are cowardly, anxious, shy, over- or under-educated, not very stable – in other words, ordinary people'. And he rightly complains that matters which closely concern us all are left to chance or to the calculations of computers, while the specialists deliberate together and publish dry reports. There are no generalists devoted to studying the continuing development of the whole man, no social educators helping him to overcome his fear, that arch-enemy of tolerance and progress. There are no 'catalysing agents' enabling him to fulfil his potential. And finally, there is no one to prepare him for his new life, help him to maturity and ensure that, when the time is right, he in his turn becomes a teacher. The normal educated person, Hauser points out, is not interested in methodology or impressed by the mysteries of technical language; but he is often desperate to bring out and express the wisdom, goodness and sense of justice hidden deep inside him. To help him to do this is education in the socratic sense.

Well, if there are people who do this work, then they are Richard Hauser and his assistants. He is tireless as a writer, lecturer, reformer, and founder and animator of new groups. He is always trying to make the speechless speak, to unite the solitary and to give them 'central heating as well as human warmth'. But chiefly he wants to teach them to ask questions. 'Why?', he says, 'is the most important word in the language.'

Hauser values practice far above theory. When I wrote to him years ago asking for written material about his work, he replied: 'Reading and writing is not enough. It's better to lend a hand.' It was good advice and I have followed it in my fashion. For me he made the first gap in the defences that separate the intellectual from those he should be working for.

Many eyes see better

In the field of community research we shall in future see, in addition to numerous existentialist group experiments, much more strenuous and financially better supported scientific inquiries. They should be given priority rating in view of the dangerously rapid increase in the population which will necessitate much practice of co-operation and mutual tolerance. Work is already in progress on crowding (Calhoun and Leyhausen), on aggression considered both biologically and psychologically, and on the composition and behaviour of a wide variety of groups. A vast field of research lies ahead.

Many voices have drawn attention to the backwardness and one-sidedness of studies of this kind which are so particularly urgent at this time. I shall quote one, the behaviourist Irenäus Eibl-Eibesfeldt:

> A great deal is talked about man and his aggression, yet until now there has been no investigation of the most elementary social behaviour. . . . For instance, European studies neglect the question of how sympathy, the urge to comfort and the readiness to be reconciled develop in children. Why is it that a third person will intervene when two are quarrelling? My own theory is that we are 'pre-programmed', and certainly not only in the field of aggression: we also have various predispositions towards goodness which could be particularly useful in research for peace. The important thing is to understand the mechanism: why is man amenable, why is he aggressive?

One of the most important insights in the last few years in the field of human co-existence has come from 'critical anthropologists' showing that the claims of previous ethnologists and anthropologists to objectivity have been quite false, since they proceed on the basis of purely Western values, attitudes and measures. Their findings have been coloured by this fact. The result has been failures in practical attempts at co-existence between white and coloured peoples because the forms adopted have depended on white misconceptions. Tension, friction and war were bound to follow, especially as the differences were exacerbated by a conflict of economic and other interests.

These critical anthropologists tell us, for instance, that the figure of the Indian chief was simply a reaction of the American Indians to their white enemies: the latter, following their own patterns, demanded a 'responsible leader' when they came to negotiate with the tribes who had originally lived in leaderless communities.

Maguroh Maruyama, a Japanese living in the US, therefore calls for a 'poly-ocular anthropology': not only Western researchers with their totally different viewpoint, logic and method should be allowed to make generally accepted statements about the life-style, views and aspirations of peoples who have had a quite different development. Instead of this false objectivity which only misleads, we should have comparative discussions.

It would be especially interesting to find out how non-Western observers see our social patterns. A first attempt has been made in the form of a video film on the 'habits and customs of the inhabitants of inner Chicago'. It was shown to both white and coloured social scientists, and has already resulted in fruitful discussions and attempts at mutual understanding.

We have recognized that the way to co-operation and co-existence lies not through playing down, veiling or suppressing differences, but through common efforts to solve conflicts. This recognition expresses itself in the fact that peace research largely takes the form of crisis and conflict research.

It is hard for the layman to regard as concrete contributions to peace the dynamic-mathematical models I was shown at the Ann Arbor Center for Peace Research and Conflict Resolution. But I am prepared to believe my guide: he said that, apart from their fundamental significance, these complicated abstracts and calculations had already had practical results 'behind the scenes'. In his view, these analyses had struck several Pentagon strategists as an X-ray of a smoker's lungs might strike a chain-smoker. It is quite possible that highly regarded institutes for peace research, especially the government-maintained research institute in Stockholm and the London Institute for Strategic Studies, may have had braking effects which the public knows nothing about. Only the memoirs of the future will be able to tell us for certain whether these are historical facts or myths spread with an eye to the budget.

In coming years the results of group and peace research will also become apparent in the increasing influence of people now being trained as peace-makers and peace-keepers of various kinds. They will be found in schools and businesses, in the political arena, and in many other situations where they will diagnose crises, examine areas of agreement and conflict, work out gradual processes for changing and improving the situation, and generally act as practised mediators and pathfinders. There have always been people who have been born with the talent to do these things: they will not be replaced, only supported by trained and practised peace-makers.

In the view of humanist psychologists, members of this important profession should be trained not only intellectually, but physically as well, in order to develop a sense of physical pressure and its consequences. Here they are in agreement with the French neo-Marxist Henri Lefebvre, who made the following thought-provoking observation in his studies on revolution and war: 'Our civilization, our culture, our society, both state capitalism and state socialism, are all based on contempt for the body, on the disappearance of the body. It is not just that the body is viewed as an object. It has gone: we break it down into various images, look at it, read about it, but we cannot feel it.'

The rediscovered body

A group of people are sitting in a meadow. In the middle of the circle lies a loaf of bread. After a short silence one of the people picks it up, weighs it in his hand, smells it, feels the crust, and finally hands it on. As soon as the loaf has made the round, one person breaks off a bit and passes on the rest to his neighbour. Each person closes his eyes as he chews his portion. The portions are not swallowed until they have completely dissolved. Everyone opens his or her eyes and looks at the others in silence.

This is one of many exercises used in the courses held at the Esalen Institute at Big Sur on the Pacific. The object is to bring atrophied senses back to life. But the ceremony is also concerned with the meaning of bread, with the act of dividing it and with the communion of eating together. The main purpose of this and hundreds of other exercises is to reawaken physical and psychological feelings that have been suppressed in our civilization with its exaggerated emphasis on the head, on the abstract, on analysis and on the intellect.

These efforts are aimed first of all at breaking down the inhibitions, habits and prohibitions that envelop man like a suit of armour. This is done by using words in an aggressive, shameless or wounding manner in order to lure a person out from behind his defences; or by undressing for games, dancing, communal bathing, mutual massage. Emotions are set free by screaming, crying, hitting, laughing; silences serve to look inwards, listen inwards, become conscious of the self.

Some of these methods strike one as very American, especially in exaggerated reports and rumours; but basically they belong to the Old World of Asia and Europe and now they are returning to their place of origin by way of the US.

In this context Germany is very important, and especially Berlin in the years before Hitler came to power. Several movements began there which were not noticed until much later. The psychoanalyst Wilhelm Reich recognized the connection between physical and psychological inhibition as a neurotic defence and one of the causes for the failure to achieve full sexual orgasm. H. J. Schultz, who is still scarcely known, developed his method of 'autogenous training'. Count Dürkheim and his students studied Japanese attitudes and drew attention to the belly region, the *hara*, as man's centre of vitality. At Humboldt University Wolfgang Köhler taught the 'Gestalt' psychology which he had developed together with Karl Bühler, Kurt Koffka and Max Wertheimer. Kurt Lewin laid the foundations of his group dynamics. J. Moreno and his pupils practised 'sociodrama' – slipping in and out of different roles and practising non-verbal methods of expression.

Then there was the musical educationalist Heinrich Jacoby. He did important work on training special talents through emphasizing the expressive possibilities of speech, movement and plastic creation. All his efforts were aimed at leading the pupil towards his task by his inner feelings. 'Be ready to experience' was one of his slogans; and his partner, the gymnast Elsa Gindler, told her pupils to open up their bodies, wake up, 'be fully there'.

It is hard to explain what exactly was done in these courses. There were certainly no strenuous physical exercises to exhaust the body purely externally. People lay on the floor with a broomstick under their spines. They tried to feel this uncomfortable object in every detail; then they would push it away and try to give themselves up entirely to the ground that bore them, to feel it holding them, to feel their whole bodies from the tips of their toes to every single fingertip, from the belly to the chest to the shoulders and right up to the head. Then they opened their eyes, took in the light as though they were seeing it for the first time, but without grabbing at it, merely holding themselves open, actively waiting for an encounter with all around them and with themselves.

As I write this, the mechanism of heightened awareness which I learnt years ago immediately begins to work again within me. Only a minute ago I did not notice how I was sitting, my fingertips moved across the typewriter keys without feeling their counter-pressure. Now I can feel a gentle breeze coming through the window. It was probably there all the time, but I had 'excluded' it. I feel my neck, my back and my feet. My big toes have begun to tingle. Of course! Waking up the feet, making them come alive, was one of Mrs Gindler's specialities. She showed us how to transform those dead lumps without a single movement, just by 'feeling through', and turn them into warm pulsating parts of our physical selves.

This kind of sensory awareness was introduced into the US during the thirties by a pupil of Mrs Gindler. It was the first impetus towards the new trend. Since then, sensuousness has beome one of the chief aims of the American counter-culture: so much so that young scientists are trying to admit direct sensory awareness and feeling back into the laboratory, from which it was previously excluded, and to use it as a complementary extension of science.

'Fools do not feel,' says an old Hebrew proverb. I found it in the writings of Moshe Feldenkrais, a physicist who, after working for years with Frédéric Joliot-Curie, turned to behavioural psychology and began to study the experiences of his own body. He regards the 'self-perception' of man as the prerequisite for reawakening his body and his mind. Working at his own institute in Tel Aviv, Feldenkrais has invented a new kind of physical self-training whose object is to wake up the coachman who has fallen asleep on the box of the coach – the coach being man himself.

His credo of the 'upright walk' points beyond the physical. But he says that 'an awareness of his organic needs forms the basis on which a person can get to know himself. If he becomes aware of the relationship linking these instincts with their origins in the beginnings of human culture, this consciousness will provide him with opportunities for directing his life which have so far been recognized by only a very few. I believe we are living in a transitional period which heralds the rise of *Homo humanus*, the whole man. He may even appear in our lifetimes.'

The rediscovery of the body can be intercepted and turned into something banal and harmless, just as the nostalgia for nature in the descendants of peasants who had moved to the towns was debased and caricatured in the form of allotments, or as the first flood of physical strength made superfluous by modern society was confined to the sports ground. But now we have seen through these diversionary tactics, and it is therefore to be expected that these impulses will be allowed to function in the wider framework for which they were intended, and that they too will contribute to changing a society that has become hostile or indifferent to man.

The first people to recognize that the age-old betrayal of the body is coming to an end were the doctors. Most doctors are political moderates or conservatives. Nevertheless, their daily experiences in the consulting room force them to arrive at a diagnosis critical of society. Diagnosis can no longer confine itself to the patient, but must consider to what degree he is the victim of social and economic conditions harmful to the general health.

Attempts to heal these conditions with medicine or other prescriptions can be only partially and temporarily successful. What is needed in every case is a systems analysis to examine the patient in relation to his surroundings. It is only by public stocktaking, examination and proposals for therapy that he can be permanently cured.

I think there is a strong chance that the protest of the medical profession about stress, pollution, neurosis and psychosis will lead to an examination of the social causes of these evils and eventually to counter-measures, even if this means infringing economic and political privileges. The man in the white coat nearly always gets a hearing, even when he uses his special language and arguments to demand almost the same thing for which 'the enemies of the existing order' are calling, in their case to closed ears and minds.

An International Red Cross study of the future of medicine has shown that doctors will be increasingly concerned with prophylactic measures to combat the harmful effects of modern industrial society. As well as restoring health they will try to protect it by preventive measures. But this means that the doctors must become much more closely associated with social–political decision-making. Prevention demands not only an extended training, but also a different concept of the doctor's profession and career.

Even today many countries are short of doctors. The shortage will get worse as they become increasingly absorbed – as they must do – in work affecting society as a whole. This shortage can only be prevented by forward thinking and planning. Possible counter-measures include the slackening of course entrance requirements and the broadening of training to cover a variety of 'health jobs'.

In this connection successful attempts have been made in China and the Soviet Union, firstly in the admission of more women to medical and health professions and secondly in more training for laymen.

The incipient reorientation can be understood as a revaluation of 'typically female' and a devaluation of 'typically male' attitudes and values. Strength, aggression, ambition, non-sensuousness are becoming less essential and less important; tenderness, empathy, nest-building and sensuousness are more highly prized. Women are particularly suited to the therapeutic professions, but hitherto they have been confined mainly to nursing. But experiences in the Eastern bloc over the last few years have shown that they are just as suited to all medical and health work. Even today there are far more women doctors than men in the Soviet Union, and but for military considerations there would be more still. The Russians are now trying to stop medicine becoming an even more purely female profession by making the examinations harder for women than for men.

But in the West there are still not nearly enough women doctors, and this should be remedied by a deliberate policy to encourage them. Entrance to the profession could be facilitated by better medical education both in schools and for adults. In China and Vietnam 'barefoot doctors' are trained for a wide range of para-medical occupations in every village and factory: this system could be imitated in the highly industrialized countries as well.

Medical information for the layman has been neglected to the point of irresponsibility – a fact that has been gleefully exploited by the popular press. It is due largely to the attitude of a section of the medical profession: they rightly fear the abuse of medical knowledge, but they are also afraid that wider medical education might undermine their authority. One of the many relics of élitism that remain in our society is the doctors' attitude to their patients: they leave them in ignorance of what they know about them and are going to do to them. But this attitude, too, is on the decline, and one hopes that with the next generation of doctors it will have disappeared.

A knowledge of the conditions favourable to disease or to its cure is a strong incentive to change one's ideas and one's actions. People used to bear the physical injuries of enforced industrialization patiently because of their ignorance, and because insufficient attention was paid to the harmful side-effects of chemicals, technology, urbanization and work stress. Efforts at enlightenment have been going on for years, and now at last those affected are beginning to listen.

It is difficult to imagine a reversal of this reawakening of the body and of the resultant 'health wave'. Even the false panaceas of altered feeling states through drugs or sedation through propaganda are not likely to be lastingly effective. The new attitude to the body must entail the breakdown of those social structures that are hostile to it.

When the skin begins to hear

Ever since the beginning of the debate about the limits of growth one question has been asked with increasing frequency: are there any less dangerous directions in which the human desire for development could move? Presumably attempts at reorientation will be directed not only towards increasing man's physical well-being, but also to increasing his sensual, intellectual and spiritual powers.

These endeavours fall into various categories. One of them has been defined by Lawrence J. Fogel of San Diego, a specialist in the study of man–machine systems. He fears that as machines become more and more sensitive, accurate and powerful, man will become more and more dependent on them until 'the intelligent use of man by his machines will become the rule'. He therefore considers it increasingly important to strengthen the human factor by constantly improving man's channels for receiving information from his environment, that is the senses of sight, hearing, feeling, taste and smell; he also wants to improve the way in which the human brain converts this input into decisions and actions, and to make the transmission of these mental processes to the outside faster, more complete, smoother and more exact.

John Platt of the University of Michigan has been thinking along the same lines: only a small fraction of the wealth of impressions, patterns, feelings and thoughts that go through our heads every second can be expressed in the written or spoken word. That is because we have too few output channels: only the mouth for speech and the hand for writing. Would it be possible to relieve these bottlenecks by finding new channels?

One possible method could be to imitate the deaf and dumb, who are able to convey information by manifold movements of their hands and fingers. Platt asks:

> Could such a system be invented? I think it could. If we look to see where we might have a set of multiple parallel muscular out-puts under conscious control, we find them in the hands. Could we not fit a light rubber glove, say to the left hand of a small child, and teach him to manipulate his muscles like a violinist or a pianist, so that several electrical contacts in the glove would pick up his muscle-signals in a precise way? The hand has 19 bones with opposed pairs of muscles, and this would give at least 19 parallel channels, whose signals could then be broadcast from a citizens'-band radio on the child's wrist. A teacher or another child with a similar radio and glove might pick up these signals and have them transformed back into electrical pulses, which could then be received as prickles, say, from 19 small electrodes on the back of the hand.
>
> With such a system I might be able to flex my fingers in a certain way and get across to you perhaps a whole graph, or a sketched picture or cartoon, or a whole sentence or a pattern of ideas, all in a single gesture, at once.

If such a language could be learnt in early childhood, then we should presumably have many thoughts and ideas which are now filtered out and never reach the surface of our consciousness because we know that we can never use them. Then we might recognize patterns which at present we have to divide up into hundreds of separate facts, because we are not capable of expressing them as a whole. And imagine what new perspectives would open up if not only Platt's nineteen lines from the hand, but other parts of the body as well could be used as transmitters and receivers. Platt speculates that this might lead us far beyond speech as we now know it, just as speech itself replaced mere inarticulate grunts. Groups of thinkers could work together with a new ease of communication and inter- action, and a whole new level of group creativity would become feasible. He holds that our present speech is probably as incapable of expressing all possibilities as the pre-verbal language of inarticulate sounds was incapable of compassing the full range of our verbal means of communication.

The skin as an organ of 'speech' has been intensively studied for many years. It was Rousseau who discovered that, if you touch a violin, you can tell by the vibration at your fingertips whether it is playing a tune in the major or minor key. Since 1962, Frank E. Geldard in his Cutaneous Communications Laboratory at Princeton has been developing the alphabet of a 'body language'; its signals can be transmitted to various parts of the body by means of vibrators, and after a little practice they can be read faster than Morse signals received through the ear.

During the first stage of his work he developed the Vibratese system, which can transmit and receive forty-five signals. They are distinguished by three degrees of intensity (weak, medium, strong) and three durations (short, medium, long); and they have five different meanings, according to which of five different specially sensitive areas of the chest they are transmitted to.

The next stage was to determine, by means of careful experiments, five further particularly sensitive reception areas. For the moment the range will not be extended, because experiments have shown that this leads to an overlap of sensations, as in a radio that can pick up a great many stations but can only transmit a limited number without interference.

Much progress in the field of body language has been made with the development of a better input mechanism, the 'optohapt'. This is attached to a typewriter, and as the keys are struck it turns them into photoelectric impulses: these are translated into vibrations which can be 'played' on different skin areas as on a piano. Describ- ing the feeling of the new language, Geldard says that some of these patterns positively force you to give them your attention. The letter W, for instance, runs right down from shoulder to ankle twice over, 'flashing like a tactile neon sign'.

At the Stanford Research Institute at Palo Alto a 'bio-information' group is working on sharpening and heightening all human sensory impressions. They explore and practise not only feeling, but also hearing, better and more precise distinction in seeing, even awareness of the circulatory functions.

It seems a strange interest for such a famous think tank, which has been working on million-dollar projects for the Navy, the Army and big business since 1946. The alleged object is better communication for the blind and deaf, but the main motive here too is armaments research. Pilots in Korea and Vietnam were often unable to receive messages intended for them because of noise, anti-aircraft searchlights and many other difficulties. So researchers began to wonder whether they might not be reached through their skins, or, quite literally, through the seats of their pants.

But here too the attempts to increase sensory perception have broken through the military framework. Scientists who refuse to go on working as handymen for the armaments industry drop out and take their experience with them to peacefully oriented research stations.

The function of the useless

In accordance with the new development, the body is no longer regarded merely as a means to increase power and achievement but also as an instrument of play. Many young people nowadays think it much more important to widen and intensify their experience than to manufacture and acquire products. They explore the neglected and underdeveloped territory of the senses in music, dance, and particularly in a newly discovered 'art of love', developing an extended and refined range of caresses without fear or guilt feelings, without violence or haste.

As he unfolds and develops, millennial man finds new meaning in play in all its forms and possibilities: play enters into spheres which used to be ruled by utility alone. In his Paris studio, Nicholas Schoeffer uses electronics and light to 'manufacture' surprise, amazement and aesthetic pleasure. Jean Tinguely creates happy, grotesque machines which, instead of producing useful objects, stamp, make shapes, twist and caricature themselves. Such things are more than aesthetic experiences: they liberate the spectator from the pressures of the technical world and its strictly prescribed goals.

Following these two pioneers, a whole generation of artists has devoted itself to having fun with wires and cables, soldering irons and welding torches, microphones and micro-circuits, light cells and computers. At a new level of development, they are acting out the dictum from the Japanese *Book of Tea*: 'The first man to pick a bunch of flowers rose above the animals. He became human, because he wanted more from nature than the bare necessities of life. When he recognized the use of useless things, he entered the world of art.'

This play, often with technical elements, looks like mere fun, but it has a hidden significance. Most people are constantly in search of the 'useful': the artists show them that man should be the inventor, creator and partner of his machines, but never their handyman or slave. In the eyes of the many they may seem mad; but really their function is to bring the world back to sanity, and that is what they do. And more: they create the necessary visions of totally different attitudes and events. That is why their happenings, actions, projects, simulation games, explosions, media mixing, optical, haptical and acoustic farces and jokes can tell us more about the 'alternative future' than the rigorously logical extrapolations and prognoses of scientific futurologists.

I first realized the prophetic function of the artist when I was standing among the ruins of Berlin shortly after the end of the Second World War. I was looking at the marble heads of the Prussian kings and emperors lying in the grass at the edge of the Siegesallee. They had been blasted off by bombs, and moss was growing from the corners of their eyes and weeds from their cracked mouths. Where had I seen it all before? Yes – only a few hundred yards and about twenty years from this very place – at a Surrealist exhibition held in an avant-garde gallery of the Tiergarten quarter during the twenties!

Artists and poets are sensitive to the future, and they confront men and society with their as yet unrealized possibilities. Take Raymond Moretti, the son of an Italian anarchist refugee who fled to France: for years and years he has been engaged on a work of art without end. He uses plexiglass, wood, metal, colours, organ notes, jazz rhythms, words from invented languages, light spectrums, to build a magic garden, a magic city that will never be finished. Picasso loved and encouraged Moretti. He looked admiringly at his work and is supposed to have asked: 'How will you ever sell this monster?' But its very unsaleability is the characteristic and prophetic mark of this constantly growing creation: the man is not producing wares for the art market, but realizing a gigantic dream full of fears, hopes, fantasies and ideas. Late one night Roger Garaudy took me to Moretti's improvised studio in a former market hall. 'This has an order', he said, 'that makes me think of a cathedral of the twenty-first century. It is dedicated to no particular creed but to the glorification of man.'

Museums are becoming a favourite place for these liberation games with the elements and with man-made materials. In a world which often seems too monotonous, too linear or too threatening, they not only preserve the testimonials of the past, but have also become a refuge for a playful imagination. The Museum of the Twentieth Century in Vienna has huge trampolines in its 'Sky Blue Co-operative': children lose their balance on them, and adults lose their acquired dignity. Wherever the German inventors of 'Babbel-plast' set up their pneumatic playgrounds, bodies begin to fly, tumble, and whizz through the air. At an exhibition some years ago in the Jewish Museum in New York, cyberneticists, artists and visitors used the instruments of modern information technology to play, invent and create. Words and signs could be turned into notes, dissonances and chords by dictating them to a 'composer' linked to a piano. Solar cells on the roof of the museum were connected to huge vibrating sheets of glass on the ground floor, and these too sounded when clouds passed over. Film pictures were transferred to the skin by means of vibrators. There was an empty white-walled room with a notice explaining that it was filled with ultrasonic waves.

If you wanted to, you could join actively and passively in a 'thermal experiment'. This was the beginning of the art of feeling. The visitor touched, in a certain given rhythm, surfaces with different degrees of warmth. A little later he received a slip of paper telling him the temperature of various parts of his body during this game.

As an indication of the new attitude expressed by artists, what impressed me most was John Baldassari's 'Cremation Piece'. He had exhibited an exciting photograph of leaping flames; it looked like an abstract graphic composition. Underneath was written:

One of several proposals to rid my life of accumulated art. With this project I will have all of my accumulated paintings cremated by a mortuary. The container of ashes will be interred inside a wall of the Jewish Museum. For the length of the show, there will be a commemorative plaque on the wall behind which the ashes are located. It is a reductive, recycling piece. I consider all these paintings a body of work in the real sense of the word. Will I save my life by losing it? Will a Phoenix arise from the ashes? Will the paintings having become dust become art materials again? I don't know, but I feel better.

Everything changes

Begin a new life. Give up what went before. Do what you really meant to. Don't be driven – drive yourself. Retire from the world. Collect yourself. Think. Travel without a preconceived destination: simply hope to find your direction. Work out your relationship with yourself and with the world. One should do these things. One ought to do them.

These are the catchwords of dreams. One hears them over and over again. They are signs of an inner tumult which has long ceased to be the privilege of the elect, the over-sensitive, the threatened. It has seized us all. If you take the trouble to listen to them, if they take the time, trust you, and begin to pour out their feelings, then you will hear confessions you would not have thought possible from taxi drivers, computer programmers, secretaries, postmen, metal workers, farmers, chambermaids. They express longings that can just about be expected from the middle classes, but not from 'the workers'.

What a mistake. The condition which the experts neatly label 'alienation' and file away under the letter A is not just a social phenomenon. It is a personal pain that torments us all and never leaves us quite in peace. It is a breeding ground for unrest, longings, fears, timid impulses and innumerable daydreams.

We have spoken a great deal about external crises and not enough of inner ones. They are hard to document, because each has its own individual development, its own highs and lows, its own alarms and appeasements, its own moments of quiescence and of breakthrough. Since I have begun to talk to 'ordinary people' whenever I have the chance to break down their shyness and my own, I have noticed that the professional 'revolutionaries' really know far too little of what moves those for whose sake they want to change the world. Each individual wants first of all to discover himself. But everyday life still runs along the old lines and cannot follow this new trend quickly enough. Those who cannot wait and try to find self-realization for themselves have a hard row to hoe: they are branded as dropouts from society and as traitors.

The eminent American historian and philosopher of science Thomas Kuhn has developed the very fruitful though simple concept of the 'paradigm', the frame of reference which continues to hold together all the single elements in the intellectual life of an epoch until more and more insights, experiences and ideas arise that conflict with the existing scheme of things and are incompatible with current views. Then quite suddenly – sometimes with a single intellectual breakthrough as in the cases of Galileo, Copernicus, Einstein or Planck – a 'paradigm shift' occurs which reconciles the contradictions in a new way. I mention this theory because I feel that at the turn of this millennium the contradictions between inner and outer demands, between data and events and between the individual and society have become unbearable and are going through a process of change. What threatens to tear us apart today may soon find a new harmony within a different framework. The catalyst of this shift is the new conception of what is and what will be, of the present and the future, of what is fixed and what is mobile.

From now on facts will be regarded as tiny parts of a constant process of change. Reality has many perpetually changing faces and is characterized by dynamic processes instead of static conditions. Precisely defined facts turn out to be no more than conventions with which only temporary agreement is possible: we know that development has passed on, that the perpetually moving film of events has long since left behind the static snapshots over which we continue to brood.

The constantly expanding universe of modern cosmology, the concepts of a moving space–time continuum that came in with nuclear physics, the biological discovery of ceaseless vital processes among the minutest living organisms – all these things have scarcely touched our historical, political and social life, which lags far behind the latest insights of the natural sciences. But now at last the new dynamic concept of the world is beginning to make itself felt on this level also.

What will it mean? A devaluation of all rigid laws, divisions and frontiers. Constant renewal, continual change, a high degree of mobility, floating – not only for currencies, but as a valid attitude in many spheres of social life. Safety lies not in clinging to the sinking wreck, but in swimming on waves that do not pull you down but carry you forwards.

Groups, however hard they try to become more open and mobile than hitherto, will hardly be able to attain the necessary degree of receptivity, so individuals will have to take on the role of fore-runners and experimenters, prepared to experience and communicate: they will play the part of sensors and antennae. But they can only do this if they regard the exploration, extension and deepening of their own individualities as a task which will not only help them to realize themselves, but will also give decisive new impulses to society as a whole. Then self-discovery will not be a flight from society, but an expedition undertaken not only from self-interest, but in the public interest as well. The process of self-discovery must include, as Reimar Lenz of Berlin has emphasized, an effort to unify our scattered knowledge and to become a 'potential planetary information centre'; we must begin to collect, work through, synthesize and experience the knowledge that we have about ourselves. But this can only be a stage on the journey of millennial man. We must try to keep up with the stream of events by striding and leaping forward, by sinking down and rising higher. Better still, we should be a stone's throw ahead.

Epilogue

It has got darker since I started to write this book. All over the world the lights of hope have grown dimmer in the recent past:

◇   Many new social projects have failed or been cancelled for lack of financial and moral support.

◇   Attempts to humanize the world of work have been abandoned because they were too expensive or inefficient.

◇   Efforts to create a new technology have been minimally successful – with a few exceptions in the field of solar energy.

◇   Anxiety about the destruction of the environment has dangerously abated as suddenly as it arose.

◇   Open people are closing up again because they have been hurt or exploited.

These failures have produced a profound sense of resignation in many who feel that efforts towards a peaceful transformation of society have definitely gone astray. Consequently some have withdrawn completely into their own private interests, while others expect disasters like war or bloody revolution because they consider them to be the only way to achieve the necessary profound changes. In this situation it is more than ever a matter of urgency to confront the counsels of despair with the counsels of hope.

In times of rollback like the present, the preparatory work for visions and designs of a humane society is more important than ever. Past defeats can be discouraging, but they can also lead to an improvement of former models whose defects have become apparent. Today the groups who have not given up the struggle are all characterized by a high degree of self-criticism, a more profound analysis of external obstacles, an increased capacity for solving problems and finding alternatives, and not least by their commitment to hope as the essential and unshakeable driving force towards change.

W. W. Harman of the Center for the Study of Social Policy, Menlo Park, California, used a great wealth of material in his major study *Changing Images of Man* (May 1974). He showed that the radical transformation of man and society at the turn of the century was unavoidable and predictable.

When I discussed the retrograde developments in the world today with him, he declared that the true innovators had certainly not given up, but they were publicizing their work much less or not at all. As a result they were less distracted by premature outside interest in their work and able to concentrate more than before. It was no more than a temporary retreat, he said, reminiscent of the early Christians in their catacombs. In this situation the following question becomes more and more urgent: is it not possible that all our past attempts to change the dangerous course of human history have been too feeble and have come too late? This is a justifiable anxiety, which must not be assuaged with unfounded words of comfort. All the more true because – contrary to occasional earlier appearances – the first signs of scarcities in the developed world did not lead to a change in demand or in our way of life. The 'energy shock' was quickly forgotten, and now we are back to 'business as usual' with inflated prices (and profits) – in spite of all the warnings we have had.

But when you look at the situation as a whole, this view too turns out to be superficial. Our situation with regard to energy and its use has not changed as dramatically as we assumed it would; on the other hand there have been a number of measures whose effects will not appear for several years. Among them are:

◇   the discovery of new sources of energy and the shift from oil to other types of fuel;

◇   the setting up of national and international bodies to examine the long-term prospects of energy supply. These include the study groups of the International Institute for Applied Systems Analysis at Schloss Laxenburg in Austria, which is attended by researchers from East and West;

◇   the ongoing discussion of these topics the world around, in the course of which the doubtful role of the big oil companies has been revealed to millions for the first time.

But so far even those who have realized that the waste of energy must be halted have often been unable to put their resolution to save it into practice, because the changeover from energy-wasting to energy-saving systems implies innumerable drastic alterations in communications, working conditions, housing, etc. These alterations will take not months, but years or even decades to accomplish.

Naturally the forces of reaction oppose these changes as they oppose all others; they passionately defend the old methods and only pretend to accept new ideas while in fact trying to sabotage them. But these are only rearguard skirmishes, even if the powers opposed to fundamental change sometimes seem invincible. After all, they are on the defensive. They are still in their positions of power, but the irreversible process of time is against them and is undermining their positions, even though this cannot be quantified.

Modern man has grown up with technical appliances, and this has bred a habit of impatience which makes it impossible for him to understand that the light of the new truth cannot be switched on as suddenly as the light in his sitting room. Changes in society, changes in our way and style of living, new types of behaviour, new ideas of culture and new scientific paradigms will bring about a radical transformation of the world: but it will come only after long, hard and determined struggles during which some of the ground that has been gained may be temporarily lost again.

The vanguard of the new always has the hardest battle to fight because it is inferior in numbers and strength to the so-called realists who recognize as real only that which is accepted and dominates the present scene. These 'realists' showed their lack of a true sense of reality when, at the turn of the century, they ridiculed women's attempts to liberate themselves and underestimated the drive of colonial 'natives' towards freedom. They hold similar views about most of the efforts described in this book, because they cannot understand that reality comprises not only that which exists and is generally recognized, but also what is still nascent or developing.

The question arises of whether a mature society should not extend the same degree of care and protection to new-born ideas as it does to its own biological offspring. If the pioneers of new ideas were treated with tolerance and sympathy instead of being opposed and maligned, if those in power were prepared to shelter and succour the first uncertain, groping, experimental developments towards something new because it might contain the answer to their own unsolved problems, then in many cases there might still be time to save ourselves.

But as things are today we shall have at best to be content with piecemeal salvation: many crises, breakdowns and disasters will inevitably claim their victims before they can be halted.

It is sad to think that we do not usually try new ways until necessity forces us. The result is that large-scale innovations start under the worst possible conditions, and that leads to many more failures and forced concessions. New ideas suffer almost more from distortion and misrepresentation than they do from premature failures, and so they come to be seen as caricatures or monsters.

But it could be that we shall learn not only to diagnose our previous failures in critical situations more accurately than before, but also to draw therapeutic consequences from them: in that case the ills before us may perhaps be cured without drastic operations. This book shows that we are beginning to develop our knowledge of great dynamic systems and sub-systems, of their anatomy and pathology and the changing conditions for their existence: this knowledge should enable us to make interventions of a carefully aimed and calculated kind.

Just as a doctor knows that he cannot cure his patient without the patient's own co-operation and will to get better, so those responsible for the survival of threatened mankind will have to make sure they have the active, creative support of as many live cells within the system as possible. The politicians and statesmen of tomorrow will be the physicians of society, and they will have to work through counsel and suggestion rather than by command and coercion. The last decision about man's fate lies with a more highly developed Everyman.

In 1955 Erich Fromm ended his book *The Sane Society* with the following passage:

> Man today is confronted with the most fundamental choice: not that between Capitalism or Communism, but that between *robotism* (of both the capitalist and the communist variety), or Humanistic Communitarian Socialism. Most facts seem to indicate that he is choosing robotism, and that means, in the long run, insanity and destruction. But all these facts are not strong enough to destroy faith in man's reason, good will and sanity. As long as we can think of other alternatives, we are not lost; as long as we can consult together and plan together, we can hope. But, indeed, the shadows are lengthening; the voices of insanity are becoming louder. We are in reach of achieving a state of humanity which corresponds to the vision of our great teachers; yet we are in danger of the destruction of all civilization, or of robotization. A small tribe was told thousands of years ago 'I put before you life and death, blessing and curse – and you chose life.' This is our choice too.

I have tried to collect experiences and material from around the world that give grounds for hoping that man will choose the right direction. But my optimism – admittedly willed – is meant to be not a sedative but a challenge. It is deliberately directed against the prevailing mood of resignation and despair, though not without concrete, verifiable arguments and phenomena.

Clear signals of the dangers ahead are multiplying; but so are the signs – albeit mostly still weak – that the crises of mankind and individual human beings may yet be overcome without invincible disasters and new tyrannies.

'Technological fixes' can at most provide short periods of remission. They cannot be expected to yield long-term solutions. The only real hope lies in a profound change in man himself, and that implies a change in society and its aims and values. Religious and ethical efforts have not brought about this change; neither have social or political revolutions, in spite of all the efforts that have been made.

But now the need for survival is forcing through a gradual human and social transformation. Millennial man will survive *and* flourish if he recognizes and seizes his chances.

PART TWO: TOOL-KIT

Suggestions, information, contacts, materials,
notes, sources, quotations

What follows is not intended as
a scholarly bibliography, but as
an effort to furnish the reader
with tools for further exploration
of the subject, for contacts with
other people and for social action
leading to change.

Perhaps it will be more amusing
to rummage through this 'tool-kit'
than to use it as a text to be read
in sequence.

Crisis Research

An inventory of the problems

The suggestion that the potentially dangerous era of the turn of the millennium would require a special investigation into possible crises – through the establishment of one or more diagnostic centres for the ills of the world – has been voiced since the beginning of this century, above all by H. G. Wells. But only rudimentary efforts have so far been made: crisis studies have been undertaken by several governments, by the specialist agencies of the United Nations, by professional associations and also by some large international industrial concerns. But there is no sign yet of a vast data-collecting network for social and economic problems to serve as a permanent global early-warning system, as suggested in an essay entitled 'A Data-Collecting Network for the Sociosphere', by Kenneth Boulding of the University of Colorado, published in *Impact* (UNESCO, Paris), April–June 1968.

The important conference on the environment in Stockholm in June 1972 decided to create a worldwide information system for environmental development. The work of an interdisciplinary research group at MIT, *Man's Impact on the Global Environment* (Cambridge, Mass., 1970), serves as a model. Numerous studies by the Club of Rome, and the vast global studies by the International Labour Office (ILO), published in Geneva in 1973, point in the same direction.

The aim is a worldwide collection of all sources of information such as has already been accomplished to a large extent in their special field by the International Atomic Energy Agency. The UNISIST group of UNESCO, under the direction of Harrison Brown of the US National Academy of Science, is hoping to put this into practice not later than 1978, in spite of numerous political as well as technical difficulties. In addition there are private projects such as the one by the group 'Mankind 2000' in Brussels which, since the beginning of 1973, has been attempting to draw up an inventory of all world problems, not only in the fields of economics, sociology, ecology and politics, but also in those of psychological and many as yet neglected difficulties. A *Yearbook of World Problems* is to be produced annually by A. Judge and J. Wellesley-Wesley which, it is hoped, will stimulate public participation in seeking solutions. The first issue was published in April 1976. Since July 1973 a steady stream of research reports devoted to systemic world problems has come from the International Institute for Applied Systems Analysis (IIASA), Schloss Laxenburg, Austria.

Social indicators

Those trying to record social indicators have discovered the difficulty of registering data which cannot be summed up in such concrete terms as can, say, quantities of steel or energy. Since the mood of crisis is due above all to a growing dissatisfaction with the 'quality of life', crisis studies will also have to concern themselves with this problem. In the years since 1965, when the social indicators movement began, hundreds of publications have appeared on this subject. Amongst the most important are:

Andrew Shonfield and Stella Shaw (eds.), *Social Indicators and Social Policy* (London, 1972): this contains a large bibliography of 174 titles;

Stanford Research Institute, *Toward Master Social Indicators* (Menlo Park, California, 1969);

J. Delors, *Les Indicateurs Sociaux* (Social indicators: Paris, 1971): due to the influential position of the author with the French government, this book has the greatest chance of turning theory into practice;

R. Bauer (ed.), *Social Indicators* (Cambridge, Mass., 1966): a collection of essays which has already played a pioneering role;

E. B. Sheldon and K. C. Land of the Russel Sage Foundation, New York, 'Social Reporting for the 1970s' in *Policy Sciences* (Amsterdam), July 1972: an essay showing the further development of these efforts but also the difficulties involved.

In France, a journal edited by B. de Jouvenel, *Analyse et Prévision* (Analysis and prediction: Paris), concerns itself particularly with social indicators, as does *Social Trends* (London), started by Claus Moser in 1971.

Councils of urgent studies

Concrete and far-reaching preliminary work has been carried out by the Councils of Urgent Studies suggested by Richard A. Cellarius and John Platt in *Science* (Washington), 25 August 1972. The number of questions these two authors considered worth examining was infinitely higher than the approximately thirty 'continuous critical problems' mentioned by H. Ozbekhan, 'Toward a General Theory of Planning' in E. Jantsch (ed.), *Prospectives of Planning* (OECD, Paris, 1969).

The greater part of their list is reproduced here in order to give an impression of the size of this urgent project:

I Physical technology and engineering (crisis-related)

1 Energy sources
Nuplexes – agro-industrial complexes
Radioactive disposal
Small portable devices: nuclear batteries . . .
Other sources and conversions
Nonfossil energy: wind, tidal, geothermal
Solar power
Liquid hydrogen fuel for vehicles
New batteries and fuel cells
Special solutions for poor countries
Efficiencies of production and utilization
Utilization of waste heat
Location of power plants: multiple functions; transmission lines
Local and global limits on power

2 Material resources
Water supply
Conservation: regional ecology; management design
Nuclear desalination . . .
Minerals
Recovery and recycling: systems analysis
Low-level extraction
Substitutions: needs in poor countries; ecological impact
Land use
Classification: multiple uses
Land resource management
Restoration and reclamation . . .

3 Structures and replacement
Replacement of structures: speed and esthetics
Special fast-building, low-cost solutions: do-it yourself; domes...

4 Transportation
Auto: new engines and fuels
Air: adequacy; convenience; safety; short take-off and landing, vertical take-off and landing; noise
Rail: passenger restoration; speed; quality
Urban mass transit: speed of construction; service and convenience
Marine: new devices for speed and economy
Novel solutions: minibuses; dial-a-bus; systems approach

5 Electronics and communications
Cheaper communications and television for developing countries...
New communications and printing methods: microlibraries...
Knowledge storage, indexing and retrieval: access for the world...
Special applications: person-to-person communications; medicine; household automation; identification and credit

6 General physical and engineering problems
Ocean resources and use
Disaster research ...

II Biotechnology

7 Population problems
Better contraceptive methods ...
Mobilities of peoples: urban–rural; unused lands; immigration
Population pressure research ...

8 Food and Famine
New grains and agriculture
Alternatives to fertilizers
Microbiological sources: food from petroleum
Food from oceans: fish farming; saline agriculture; ocean farming
Genetic copying of animals for higher yields
Novel sources: digestion of grass; new biology; systems approach ...

9 *Environmental problems*
Ecological control: improved methods in agriculture; fishing; hunting
Ecological education and philosophy: resource-conserving farms and cities . . .

10 *Health – basic research*
Microbiology . . .
Aging
Neurosciences, biopsychology and behavior
Optimum environment: crowding; change; artificiality . . .
Biotechnological forecasting
Medical education: world health education and services; large-scale methods

11 *Health – therapy*
Disease research and cure: cancer; heart-stroke; neurological; aging
Low-level diagnosis; systems approach; continuous health optimization
Artificial organs and transplants
Psychopharmacology and drugs: long-term effects; poisons
Psychiatry and mental health: sanatoriums; new therapies
Emergency medical care
Nutrition: measurements; prenatal and infant
Public health: mass methods; health care delivery; hospitals; urban–rural

III Behavior and personal relations

12 *Behavioral research*
Behavior modification research: social and political effects; ethics . . .
Child development and training: early enrichment . . .
Behavior change and learning with interpersonal games

13 *Education*
Classroom teaching: new materials and methods; class management
Programmed instruction and computer-assisted learning
Educational testing . . .
Universities: structures; communities; education for careers and change

14 *Small groups*
Methods of responsive living
Family and neighbor relationships: new community housing and institutions
Group interactions: schools; churches; small businesses . . .
Group-living experiments: religious; behavior-theory; group economics and law
Child-care communities: slums; suburbs; housing and organization
Theory and philosophy of individual–group relations, emotional health
Special new roles: confidant; economic adviser; group therapist; ombudsman

IV National social structures

15 *Economics*
Inflation: removal without unemployment . . .
Aids to urban restructuring: Urban Development Corps . . .
Aids and incentives for the poor: guaranteed income; negative tax
Large-scale, long-range analysis and theory: simulation; normative

16 *Organizations*
New management methods: . . . democratization
Improved information handling and decision making
Participatory problems and humanization of organizations

17 *Mass communications*
Press: networks; control; reporter rights; privacy; underground press
Radio–television: cable television; National Educational Television; violence; public feedback; children's programs; news
New media: . . . neighborhood and small group publications
Mechanisms for increased diversity and freedom . . .
Mass communications as community and world education
Role in images of change and future; amplification of crises
Theory of effects: systems analysis; forecasting change

18 Politics
Improvement in public administration and management
Responsiveness: ombudsmen; participation; elites and checks
on them . . .
Party structures: . . . minority control; instabilities . . .
Reduction of community hostilities . . .
Education for tolerance and democracy
Information handling before and during crises
Mediation and crisis management . . .
Mechanisms of stability and change
Constitutional redesign: . . . welfare
Systems analysis and theory: long-range planning

19 Urban and rural problems
Structures: housing; streets; transport; zoning; planning
Inflow–outflow: people; food; water; garbage and sewage;
communications
Creation of new cities: regional planning . . .
Rural and farm problems: migration; economics; quality of
life . . .
Esthetic and cultural requirements: funding; cultural diversity

20 Large-scale change
Population pressures: mitigation; use patterns; redistribution
incentives
Quality of life: recreation; esthetics; differentiation; minorities
Social indicators
Systems analysis: forecasting; theory of change; megalopolis–
ecumenopolis . . .
Systems analysis of social patterns with rewards instead of
punishments . . .
Reward systems for social inventions and improvements

V World structure

21 Peace-keeping structure
Contingency plans for possible new peace-keeping mechanisms
Local war mediation and control methods: arms reduction
Military–industrial lock-ins to policy: conversion to new roles
Systems analysis of alternative world structures

22 Economic development
Mechanisms of investment and growth
International monetary stabilization . . .
Transition to steady-state consumption: development . . .
Unemployment

23 Developing countries
Education, large-scale: local and world language; television
Easing of change: preservation of values; independence
Damping of racial and national hostilities: education; commer-
cial payoffs
Government and political restructuring
Mechanisms of change
Pressures of technology
Education for democratic management . . ..

VI Channels of effectiveness

24 Political and economic support of urgent research
Case studies of social innovations . . .
Organization of interdisciplinary centres for urgent studies . . .
Technical advisory services to legislatures, industry, and public
groups
Self-supporting research developments: new businesses and
industries
Contacts and education for broad support . . .

25 Systems analysis
Mapping of problem areas and studies: resources; progress;
feedback
Theory of organization, structure, and growth of urgent
studies
Large-scale, long-range systems analysis: world dynamics;
hierarchical jumps; the global system; ecumenopolis
Match of new innovations to long-range directions and self-
determination
Democratic theory of group and social choices, and checks
and balances in the process of complex change
Philosophical structure integrating these changes and studies:
long-range evolutionary; normative; personal-behavioral;
human benefit and self-fulfillment

Creativity

General survey

An excellent collection of twenty-seven essays and personal accounts is provided by P. E. Vernon (ed.), *Creativity* (London, 1972). The history and theory of the concept, personal creativity and methods for stimulating creative thinking are most competently discussed in this book.

Comprehensive bibliographies on the subject of creativity research are published regularly in *Journal of Creative Behavior* (Buffalo, N.Y.).

Insufficient attention is being paid to the growing contribution by French researchers and practitioners to the literature on creativity. They introduce completely new – creative! – views into a subject dominated by the Anglo-Saxon style of thinking. Henri Laborit, *L'homme imaginant* (Imaginative man: Paris, 1970) is a contribution by a doctor and biologist who stresses that 'if man sets his hopes solely on changes in his social and economic environment, however indispensable these may be, he will only partially solve the problem of alienation. . . . From now on mankind must look not only at the outside world but also at itself.' The same author's *La nouvelle grille* (The new grid: Paris, 1974) contains the following important statement:

> All progress made by man since the beginning of time has been the result of creativity. But creativity, unfortunately, has remained until now the domain of a privileged minority, set apart mainly by reason of their birth. . . . And precisely because the rarity of these creative individuals is regrettable, it is necessary to emphasize that the life of each man and woman could be creative if society would provide a suitable framework for the flowering of the imaginative abilities.

Further examples are:

L. Astruc, *Creativité et sciences humaines* (Human science and creativity: Paris, 1970);

G. Bachelard, *L'intuition de l'instant* (Intuition of the moment: Paris, 1966);

M. Demarest and M. Druel, *Psycho-pédagogie de l'invention* (Psycho-pedagogics of invention: Paris, 1970);

A. Moles, *La création scientifique* (Scientific creativeness: Paris, 1966).

Creativity research in the East

Unfortunately, I have been able to find out very little about creativity research in the communist world. At my suggestion, the subject of 'creativity of the masses' was put into the programme of the third international futurology conference in Bucharest in 1972; but no significant contributions were made by participants from communist countries.

Acta Psychologica Sinica (Peking) for August 1959 points to intensive studies in China on *ch'uang-tsao-hsing* (inventiveness): 'The psychologists, having examined the laws of creative thinking, can now contradict the heresy of bourgeois psychology which sees the gift of creativity and invention as the result of chance inspiration and sudden insight. . . .' In the same periodical for September 1959 there appears the following statement:

> The results of studies carried out by ergonomists have demolished the bourgeois misconceptions about creativity and invention. These spiritualist psychologists had maintained that creativity and invention were the privileges of a few geniuses and that ideas were the result of inspiration which came to one but did not have to be looked for.

In 1966 the Chinese press pointed out that workers had made numerous inventions in the field of their own production and research. One carpenter is reported to have suggested not less than ninety-seven innovations.

Creative personalities

Several magazine articles have been written about Leo Szilard. The one which does him most justice is contained in John Platt, *The Step to Man* (New York, 1966). His own book of scientific futuristic parables, *The Voice of the Dolphins* (New York, 1961), gives the non-scientific reader a clearer glimpse of his original way of thinking than the *Collected Scientific Papers* (Cambridge, Mass., 1972). But the best of Szilard was always spoken and not written.

On Adrien Turel in relation to creativity I recommend his autobiography *Ecce Homo* (Zurich, 1963), and *Geschichte unserer Zukunft* (A history of our future: Zurich, 1963).

It is so difficult to conduct any kind of personal relationship with Fritz Zwicky that he himself tells us that someone has suggested the introduction of a new unit of measurement, the 'Zwicky', for measuring surliness of character. In his books he often appears violent, exaggeratedly unjust and conceited; but they are full of ideas which stimulate, even after several readings. They are:

Morphologische Forschung (Morphological research: Winterthur, 1959);

Entdecken, Erfinden, Forschen im morphologischen Weltbild (Discovery, invention, research in the morphological world view: Munich, 1971);

Jeder ein Genie (Everyone a genius: Berlin, 1971).

From the last-named comes the following quotation:

> The individuality of each person is so pronounced that one could almost say that every human being is potentially a genius. It can indeed be proved in principle and by experience that every single person is capable of mental and physical achievements which no one else can emulate. My own experiences in this matter are as follows: firstly, only very few people are convinced that they are potential geniuses such as we have described them; secondly, even fewer people realize where their genius lies; thirdly, only a very few have succeeded in the course of history in recognizing their own genius, developing it and living it to the full.

One of the main rules which Zwicky advises that his students should follow is: 'No negation without subsequent constructive effort.'

In his efforts to further divergent thinking, Arthur Koestler has always paid special attention to the subject of creativity. His book *The Act of Creation* (London, 1964) is of special importance as a standard work by a non-scientist who nevertheless in this as in other instances presents his own scientific ideas, which prove to be not only informative but also inspiring. Koestler has discussed his own creative processes in an interesting volume of interviews which also contains revealing conversations with other writers and artists: S. Rosner and L. E. Abt (eds.), *The Creative Experience* (New York, 1970). Among those who here state how they 'get ideas' are Harbour Shapley (astronomer), H. Bentley Glass (geneticist), Noam Chomsky (linguist), Wilder Penfield (neurosurgeon, physician), Aaron

Copland (composer), Edward Steichen (photographer), Neil Simon (playwright).

A wealth of information is also contained in the detailed report of talks and discussions held at a conference of the 'Centre Culturel International de Cérisy-la-Salle', near Paris, in September 1970: *Art et science de la créativité* (The art and science of creativity: Paris, 1972).

Environmental conditions for creativity

Apart from Vernon's survey mentioned above, the most impressive collection of essays about creativity in the English-speaking world is H. H. Anderson (ed.), *Creativity and its Cultivation* (New York, 1959). In one of these essays the American psychiatrist Carl Rogers writes about the not uncommon case of scientists and artists who are ahead of their time in their thinking and, in their conflict with a world which feels insecure as a result, are treated as heretics or even, in certain circumstances, have to suffer death for their divergent attitudes.

The American molecular geneticist Gunther S. Stent has further developed this theme in his article 'Prematurity and Uniqueness in Scientific Discovery' in *American Scientist* (New Haven, Conn.), December 1972. He presents the idea that modes of thinking born ahead of their time remain unaccepted because they cannot be fitted into the existing mental structures. The pioneer remains out of hearing, and his voice will only be heard when the others have caught up with him.

The question of whether the 'outside world' should not change its attitude towards the new and apparently incredible and its originator is likely to remain a constant theme for debate. Too often the mediocre has been thought original by the living whilst the really new has been considered wrong or even mad. It is still so today. Fairly recently Professor F., a highly respected physicist attached to the CERN atomic energy research centre, was forcibly removed to a Geneva psychiatric clinic at the instigation of his worried colleagues, who thought his ideas 'insane'. He was not released for three days. His state of mind is now considered 'normal' and he gives regular lectures. He has learned to keep silent. This case must act as a drop of gall in the pure wine of the next title, of which I have made much use in writing this book: K. W. Deutsch, John R. Platt, D. Senghaas, *Major Advances in Social Science since 1900: An Analysis of Conditions and Effects of Creativity* (Ann Arbor, Michigan, 1970); for the authors only discuss those innovations which have succeeded and not those which have been hindered or altogether prevented.

As early as 1964, H. Klages of the Technical University in Berlin wrote a book, to which insufficient attention has been paid, about methods of innovation in modern large-scale research: *Rationalität und Spontaneität* (Rationality and spontaneity: Bielefeld, 1964). It contains an extensive bibliography.

Another comprehensive collection of essays is M. A. Coler (ed.), *Essays on Creativity in the Sciences* (New York, 1963). At Coler's institute (1 Fifth Avenue, New York) I found the largest of all institutional libraries on the subject of creativity.

Donald Schon, in *Technology and Change – The New Heraclitus* (New York, 1967), points to the obstacles which industry in particular puts in the way of the innovator. In an article 'No Room for the Searcher' for the progressive magazine *Innovation* (New York) No. 11, 1970, Robert Young discussed why especially the large firms with their cumbersome machinery and weight of responsibility have a tendency not to permit too great risks and therefore cannot tolerate the unsettling influence of the innovator and radical inventor in their midst for too long. In the previous year the same magazine published a paper by Warren Brodey entitled 'Building a Creative Environment', in which the well-known cybernetics expert suggested that research centres should be devised and built which would permit the searcher to 'play with ideas'. The laboratories built in the shoe-box style of the new functionalism are certainly not suited to this purpose.

V. A. Thompson, in *Bureaucracy and Innovation* (University of Alabama Press, 1970), mounts a sharp attack against the powerful élite of state and industry which is opposed to creativity. He calls them 'econologians' because they always measure innovation with the economic standards of the present instead of taking its future possibilities into account.

The most thorough study of the relationship between innovation and industry appears in the comprehensive report *The Rate and Direction of Inventive Activity: Economic and Social Factors* (Princeton University Press, 1962).

The fact that the whole problem is primarily political, because today's politicians either neglect or reject the establishment of an environment favourable to creativity, is developed by Georg Picht in an essay entitled 'Kreativität und Bildung' (creativity and education) in the collection *Zukunft aus Kreativität* (A future through creativity: Düsseldorf, 1971):

Pedagogics concerned with creative education is caught under the wheels of a social mechanism which expects the schools to programme young people *correctly*. . . . The creative capabilities of our children and grandchildren depend on the political creativity of the present generation.

Further references to the decisive role played by education in stimulating creativity can be found under the heading 'Creative Education' (p. 231).

An influential American educator, having left the Kennedy administration, stated his view about the necessity of a constantly self-renewing society which does not cast out the innovator but recognizes him as a most valuable fellow-citizen, in the pamphlet John W. Gardner, *Self-Renewal* (New York, 1963). To conclude this summary, here is a quotation from this pamphlet:

A society whose maturing consists simply of acquiring more firmly established ways of doing things is headed for the graveyard – even if it learns to do these things with greater and greater skill. *In the ever-renewing society what matures is a system or framework within which continuous innovation, renewal and rebirth can occur.*

Our thinking about growth and decay is dominated by the image of a single life-span, animal or vegetable. Seedling, full flower and death. 'The flower that once has blown forever dies.' But for an ever-renewing society the appropriate image is a total garden, a balanced aquarium or other ecological system. Some things are being born, other things are flourishing, still other things are dying – but the system lives on.

Over the centuries the classic question of social reform has been, 'How can we cure this or that specifiable ill?' Now we must ask another kind of question: 'How can we design a system that will continuously reform (i.e., renew) itself, beginning with presently specifiable ills and moving on to ills that we cannot now foresee?'

Social Imagination

From Utopias to blueprints

John Platt, in his crisis studies already mentioned, repeatedly points to the decisive role of 'social imagination' which had been neglected over a long period in favour of scientific and technological imagination. Under the challenge of mankind's present perilous situation it should henceforth develop rapidly.

Utopian literature – which I cannot discuss here due to its vastness – has more far-reaching but less concrete aims than social imagination. The latter can be seen as the originator of desirable and, in the not too distant future, realizable projects – 'concrete Utopias', as Ernst Bloch calls them.

This is where my own active interest in futurology began, stimulated through an article by Dennis Gabor in *Encounter* (London) 1958, called 'Inventing the Future'. From among my own numerous articles on this subject I should like to mention:

R. Jungk, 'Plädoyer für die soziale Phantasie' (A plea for social imagination) in the series *Modelle für eine neue Welt* (Models for a new world: Munich, 1963–9);

——'Imagination and the Future' in *International Social Science Bulletin* (UNESCO, Paris) No. 4, 1969;

——'The Role of Imagination in Future Research' in *Challenges from the Future* (Tokyo, 1970, 6 vols).

Later I came across the two volumes by the Dutchman Fred Polak, *The Image of the Future* (New York, 1961). He started working on this book in 1951; it shows how society's conceptions of its future influence its present daily life. Elsie Boulding of the University of Colorado has described Polak's position as a critic in an important study prepared for the 'Symposium on Cultural Futurology' of the American Anthropological Society in November 1970, *Futurology and the Imaging Capacity of the West* (Minnesota University Press, 1971). She says that, after describing how daring temporal leaps had

led to the peaks of the Renaissance, the Enlightenment and the early
industrial era, Polak turned angrily to the present and held up a
mirror to mid-twentieth-century man, showing him as the prisoner
of the moment, clinging desperately to today for fear of what tomorrow might bring. He was angry because he saw his fellow-men
neglecting a talent which they possessed, but might lose if they left it
uncultivated. The refusal to use the imagination to create other,
better futures would lead to infinite projections of present trends and
an inferior development of technological possibilities which would
eventually leave man stunted and deformed.

From among a large number of works dealing with finding alternative futures – this ungrammatical plural is used in this context in
order to differentiate between the variety of possibilities and pre-
programmed total unity of the future – the following are worth
mentioning:

O. K. Flechtheim, *Futurologie* (Cologne, 1970): this book, as well as
   Gabor, inspired me to embark on my own work in this field;
Y. Dror, 'Alternative Futures and Present Action' in his book
   Ventures in Policy Sciences (New York, 1971);
J. Galtung, *Pluralism and the Future of Human Society* (Tokyo, 1970);
Roger Garaudy, *Esthétique et Invention du Futur* (Aesthetics and
   invention of the future: Paris, 1968);
——*L'alternative* (Paris, 1972);
H. Klages, 'Measuring Social Innovations' in *Sociologia Internationalis*
   (Berlin) Vol. 1, 1970;
D. Gabor, *Innovations – Scientific, Technological and Social* (London,
   1970);
D. Stuart Conger, *Social Inventions* (Prince Albert, Saskatchewan,
   1974).
Some titles on concrete Utopian planning:
Nigel Calder, *The Environment Game* (London, 1967);
I. Illich, *Deschooling Society* (New York, 1971);
R. Schwendter, *Modelle der Radikaldemokratie* (Models for a radical
   democracy: Wuppertal, 1970);
and, above all, the somewhat frightening visions of Japanese
futurologists: Y. Masuda, *Computopia* (Tokyo, 1972), and K. Tateisi,
M. Yamamoto, I. Kon, *Sinic Theory* (Tokyo, 1970). The latter is
one of the works on 'multi-channel society', a term used by the
Association of Japanese Futurologists to describe the society of
tomorrow. It is primarily of interest because it develops ideas about a
post-technological society.

Living the future

It had been my intention to give here a survey of the many attempts at 'living the future': living communes, extended family units, experiments in industry, etc. These experiments are, however, mostly so short-lived that they would already have been out of date at the time of publication. There is also a lack of reliable documentation. In this connection a yearbook for social experiments – as suggested in the main text – would be of the utmost importance.

First hints in this direction are contained in Stan Windass, *Alternative Societies* (Oxford, 1973, mimeographed); and in a periodical edited by Dick Fairfield, *The Modern Utopian* (San Francisco), with its accompanying newsletters. This lists addresses and news not only of American communes, but also of British, Dutch and German ones.

It is also worth mentioning the literature about the Dutch 'Provos' and their various social experiments. Unfortunately, this has so far only emanated from uncritical enthusiasts or unfair critics.

Methods and application of social experimentation are dealt with by G. W. Fairweather in *Methods for Experimental Social Innovation* (New York, 1967). He suggests a plan for starting a 'Prospective Center for Experimentation', but this has apparently not been realized.

David H. Kershaw reports in *Scientific American* (New York), October 1972, on an experiment with maintenance subsidies in New Jersey, and there is an official report, *Income Maintenance Experiments*, by the US Commission on Finance, Washington.

The following works which take a critical look at society are recommended:

H. Bussiek (ed.), *Veränderung der Gesellschaft, Sechs konkrete Utopien* (Changes in society, six concrete Utopias: Frankfurt, 1970), with contributions by M. Markovic (on society), I. Fetscher (working conditions), V. Gerhardt (education), K. H. Bönner (sexuality), H. Maurer (housing), S. Quensel (crime);

R. Theobald, *An Alternative Future for America* (Chicago, 1969).

Simulations

War games

War is the father of the kind of 'serious games' with the help of which complex interrelated situations can be recognized and their future possibilities studied. Only at a fairly late stage did it become evident that these 'games' could provide valuable information about other, not easily recognizable dynamic systems, such as society or history.

The military correspondent of the London *Observer*, Andrew Wilson, gives an excellent survey in *The Bomb and the Computer* (London, 1968) of the development of military simulations, from the Prussian 'war game', thought to have been invented by a Berliner called von Reisswitz at the beginning of the nineteenth century, to the 'Joint War Games Agency', a special branch of the US General Staff.

Those interested in the details of war games since their harmless beginnings in computers will find details as fascinating as – with a little use of the imagination – they are gruesome in:

H. Kahn, *On Thermonuclear War* (Princeton, 1961);

M. G. Weiner, *An Introduction to War Games* (RAND Corp., Santa Monica, 1959);

——*War Gaming Methodology* (RAND Corp., Santa Monica, 1959);

N. C. Dalkey, *Simulations and War Games in Computers and the Policy Making Community* (Englewood Cliffs, N.J., 1968).

Systems theory

The application of simulations to civilian problems would hardly have taken on such proportions if two developments of the forties had not opened up the way for a new 'holistic' way of looking at these problems which, together with the development from the static to the dynamic view discussed in Chapter 5, has fundamentally changed our concept of the world. To cybernetics and the closely related development of computers must be added Ludwig von Bertalanffy's *General System Theory* (New York, 1968), which the author, originally a biologist, began to develop at the beginning of the twenties in Vienna. In the forties he formulated the first outline of his theory, which is based on the very ancient adage that 'everything is connected with everything else'.

With the growing specialization of research and the increasing atomization of life, mankind could no longer see the wood for the trees. Systems theory, though usually presented by its followers – with the exception of its modest and sincere inventor – in the jargon of a secret science, is in fact a perfectly simple and obvious rediscovery. It confronts the old one-sided, narrow and specialized 'scientific method' with reality in its whole complexity and turbulence. Whoever acts on this reality must develop a more detailed knowledge and understanding of interconnections than has so far been the case. Hasan Ozbekhan points to a topical example from world politics in his essay 'Planning and Human Action' in Paul Weiss (ed.), *Hierarchically Organized Systems* (New York, 1971). The worldwide hunger crisis expected towards the end of the century, he says, is a problem which cannot simply be solved by the production of more food. Neither can it be prevented by the improvement of agricultural methods. The concept of 'agriculture' as it has been understood until now is no longer sufficient: agricultural solutions alone are not enough. They by-pass the problem. For what we call the 'world hunger crisis' is connected with health, education, institutions, the size of the population, science and technology, and demands an investigation of the production and defence systems, the mechanisms of the market and of prices, and so on.

The same considerations apply if, instead of hunger, we choose another problem: education, politics, existing economic structures . . . the problems converge in a great system of difficulties. We are dealing with a situation of unprecedented dimensions and immense complexity.

His remarks apply in the first instance to very large systems; they are also valid, however, for medium-sized complexes or even smaller ones such as towns, firms, schools and families. Each can only be recognized correctly, and the prospects for its further development anticipated, by the 'total view' method.

The development of serious games

Such interrelationships are discovered by means of these games. One must distinguish between 'man games' and 'computer games'.

The first generally accessible survey, largely of 'man games', was Clark Abt, *Serious Games* (New York, 1970). Their application in connection with urban problems is outlined by Peter House, *Environmetrics* (Washington, 1968).

A political variant of these games as practised today in many foreign ministries and diplomatic schools is described by H. Goldhamer and H. Speier of the RAND Corporation in 'Some Observations on Political Gaming' in *World Politics* (London, 1959), and above all by H. Guetzkow in *Simulation in International Relations* (Englewood Cliffs, N.J., 1963) and *Simulation in Social Science* (Englewood Cliffs, N.J., 1962).

Educational, instructional and social 'psycho games' are becoming increasingly popular. They are intended to train the players in empathy and in the recognition of relationships. Their popularity may, however, be partially explained by the fact that many people have the desire to free themselves by means of these games from their static social situation or imprisonment in their job or the system.

A warning against the 'new Utopians'

The problematic nature of simulations as a possible instrument of 'domination of people by people' is the subject of Robert Boguslaw, *The New Utopians* (Englewood Cliffs, N.J., 1965). Boguslaw worked for a long time with the Systems Development Corporation, a sister organization of the RAND Corporation in Santa Monica, carrying out a variety of computer games until he came to ask himself: for whom and for what? In his book, which was written after changing to a university career, he points to the dependence of the 'new Utopians' on:

◇   the client who places the order for a simulation and knows how to influence it through his own wishes;

◇   the designers of systems and programmers who want to achieve clear results and definite decisions, even in cases where the equivocal nature of the subject does not permit this;

◇   the manufacturers of computers who are in a position to influence the data which the computer collects and processes through its design and component parts.

He pleads for a democratization of simulations. The public should be able to exert an influence and learn to understand the technology which is at present kept 'veiled'.

A similar attitude is taken by one of the pioneers of data processing in the US, E. C. Berkeley, whose critical journal *Computers and Society* (New York) has so far been able to withstand all pressures. To this may be added the doubts already voiced by Karl Popper, Karl Deutsch and others about the impossibility of expressing 'qualities' in mathematical symbols.

John Wilkinson, *Retrospective Futurology* (Santa Barbara, 1972), maintains that it is possible to overcome this difficulty with the help of methodical programming, a method developed by Professor Bellman of the University of Southern California. An expanded version of the above is J. Wilkinson, R. Bellman and R. Garaudy, *The Dynamic Programming of Human Systems* (Santa Barbara, 1973).

Meanwhile, doubts still predominate, as expressed eloquently and convincingly by Georg Picht in his essay 'Die Bedingungen des Überlebens – von den Grenzen der Meadows-Studie' (Conditions for survival – on the limitations of the Meadows project) in the monthly *Merkur* (Stuttgart), March 1973. Owing to its importance, I quote extensively:

Why have all those parameters been left out of the MIT project which refer to human thinking and consciousness and to internal changes in society? The answer is simple: they cannot be quantified, in spite of all attempts by an allegedly empirical social science to convince us of the opposite. The future development of science and technology is as incalculable as the development of political consciousness of the industrial nations and the Third World. Even a minimum of historical knowledge will teach us that the actual progress of history always depends on factors which cannot be foreseen. Adolf Hitler could not be calculated in advance, any more than could Einstein. But the world in which we live is dominated by a superstition to which politicians as well as scientists have fallen victim: the superstition that only that is real which can be quantified. All the weaknesses of the systems analysis of the Meadows project have their origins in this superstition. This statement has a tragic aspect, for it shows that the project continues along the same wrong path which led us into the very cul-de-sac which it tries to expose. Those who are led by the project to ask themselves why the evolution of mankind has been pushed in such a catastrophic direction will come to the conclusion that the triumph of science and technology, the success of which caused the Meadows trends to be set in motion, is due to the one-sided narrowness with which the human mind has concentrated on quantifying analyses. This one-sidedness has caused that asymmetry in the evolution of man which I mentioned in the beginning. It has been the cause of the backwardness of our moral and political consciousness, which lies at the centre of the dangers which threaten mankind at this moment. . . .

Even the attempts to improve the model are usually confined to trying, rather like children playing about with bricks, to see which additional parameters could be introduced. It can be predicted that those parameters will take preference which allow quantifiable results. This merely helps to reinforce the misorientation of the entire project. A systems structure in its totality and the structure of its corresponding model can only be described in a meta-language and not in quantifying statements. But all quantitative assertions inside the model are dependent on its systems structure. One of the most important conditions for the survival of mankind will therefore be the ability to free ourselves from the dominance of models which are based on the same methodological mistakes which have led us astray up to now.

The humanistic criticism of the Meadows model by the psychologist Marie Jahoda also points in the same direction: Marie Jahoda, 'Postscript on Social Change' in *Futures* (Guildford, Surrey), April 1973.

Citizens' games

Robin Roy of the Open University makes some practical suggestions for the involvement of the public in simulations of public authorities in a paper prepared for a conference on 'Assessment and the Quality of Life' at Salzburg in September 1972 under the title *Assessing the Impact of Future Socio-technical Systems by Simulation*. At the Design Research Laboratory of the University of Manchester, he tested four different new transport systems with the participation of future users by sending married couples in each of the new vehicles on a 'journey' simulated with the help of slides. Waiting at the simulated bus station, the duration of the journey, noise, obstacles, various unexpected incidents on this trip were acted out realistically.

It is possible that the 'acting out' of various future decisions by collaboration between experienced actors and members of the public might become an informative as well as an interesting form of democratic participatory leisure activity.

Alternative Schools

The growing interest in education

No other section of my tool-box for further study is crammed as full as the section on education. I am sitting here in a corner of my study like the sorcerer's apprentice trying to escape from the flood: there are books, brochures, newspaper cuttings, magazines, all on the subject of schools. How can I cope with this surfeit without boring the reader or giving him or her the feeling that I merely want to show off all this wealth of information?

This sudden worldwide interest in educational matters is surprising. The fascination which education holds has grown in leaps and bounds with the same suddenness as the interest in technology after the war. This could be significant. Or is it just a fashion?

It may be that this intensive desire for knowledge is based on a feeling, expressed by the British anthropologist Edmund Leach and his American colleague Margaret Mead, that the young know more about these exciting and critical times than we do and that they are therefore becoming our teachers. The interest in the problems and experiments in education may therefore represent an interest in young people and the prefiguration (Mead) of the future as expressed in their behaviour.

The book which helped most to arouse my interest in the formerly dusty subject of schooling was A. S. Neill, *Summerhill* (London, 1962), as well as the subsequent publications of this pioneer of anti-authoritarian education. Among numerous publications about the Summerhill experiment I would like to mention an essay by E. Bernstein in *Psychology Today* (Del Mar, California), October 1968, which describes the subsequent lives of former pupils of Summerhill and of other experimental schools.

Radical education

It is often forgotten that the roots of an anti-authoritarian school system lie in the pre-Stalinist Soviet Union, in the Weimar Republic and in the first Austrian Republic of the twenties. Light is thrown on this by Vera Schmidt, *Psychoanalytische Erziehung in Sowjetrussland* (Psychoanalytical education in Soviet Russia), a report on a 'children's home laboratory' in Moscow published in Leipzig in 1924 and piratically reprinted in 1968; and J. R. Schmid, *Freiheitspädagogik – Schulreform und Schulrevolution in Deutschland 1919–33* (Liberating education – school reform and school revolution in Germany 1919–33: Hamburg, 1973).

Further noteworthy material from among the vast number of publications on the subject of radical education:

USA
Paul Goodman, *Growing Up Absurd* (New York, 1960);
Ronald and Beatrice Gross, *Radical School Reform* (London, 1971);
John Holt, *How Children Fail* (New York, 1964) and *How Children Learn* (London, 1968);
Ivan Illich, *Deschooling Society* (New York, 1971);
Herbert R. Kohl, *The Open Class Room* (New York, 1969);
Jonathan Kozol, *Death at an Early Age* (New York, 1968);
Neil Postman and Charles Weingartner, *Teaching as a Subversive Activity* (1969);
E. Reimer, *School is Dead* (New York, 1971).
J. Rothschild and Susan Wolf, *The Children of the Counterculture* (New York, 1976).

Britain
P. and J. Ritter, *The Free Family* (London, 1959).

France
J. Celma, 'Journal d'un éducateur' (Diary of an educator) in 'Ecole émancipée' group, *La repression dans l'enseignement* (Repression in education: Paris, 1972).

Norway
M. Jorgensen, *Schuldemokratie – keine Utopie. Das Versuchsgymnasium in Oslo* (School democracy – not a Utopia. The experimental grammar school in Oslo: Hamburg, 1973).

West Germany
A. J. Breitenreicher *et al.*, *Kinderläden* (Children's co-operatives: Hamburg, 1972);

G. Fischer and collaborators, *Gesellschaft und Politik* (Society and politics: Stuttgart, 1971);

H. J. Gamm, *Kritische Schule – eine Streitschrift für die Emanzipation von Schülern und Lehrern* (The critical school – a polemical treatise for the emancipation of pupils and teachers: Munich, 1970).

East Germany
H. Stolz, A. Hermann, W. Müller, *Beiträge zur sozialistischen Erziehung* (Contributions to a socialist education).

Creative education

In the magazine *Where* (London), children were invited to make suggestions for ways of helping to prevent hostility between dogs and cats. Edward de Bono describes the results in *Children solve Problems* (London, 1972): he comes to the conclusion that the demonstrable falling-off in the capacity for creative thinking from childhood can be directly traced to our examination-based education system, with its emphasis on the passing on of accepted facts, which affords little opportunity for questioning or originality. De Bono has started a foundation, the Cognitive Trust, with the aim of spreading his ideas about thought stimulation through play and the training of teachers to awaken creativity rather than kill it.

The aim to train creative teachers is also the subject of a book with numerous examples by L. and V. Logan of Brandon University, Manitoba: *Design for Creative Teaching* (Toronto, 1971).

The question of creativity and its preservation and expansion through education is discussed by seventeen authors in Jerome Kagan, Harvard University (ed.), *Creativity and Learning* (Boston, 1968). One of the researchers represented in this collection can be regarded as the father of creative education in the US. From amongst his numerous books and publications, the best general survey is provided by E. P. Torrance, *A New Movement in Education: Creative Development* (Boston, 1965).

Ideas about how teachers can produce a 'creative climate' in schools are developed by F. E. Williams, *Classroom Ideas for Encouraging Thinking and Feeling* (Buffalo, N.Y., 1970), and M. B. Miles, *The Development of Innovative Climates in Educational Organizations* (Stanford Research Institute, Menlo Park, California, April 1969). E. Landau, *On Developing Creative Behaviour* (Tel Aviv, 1969), reports on her creative working groups with children.

Education and the Future

Equal opportunities

The interest in the future of education, and in education for the future, has grown so much in recent years that in English-speaking countries alone the number of books and essays on the subject had reached approximately eight hundred by the end of 1971. These are the findings of a bibliographical study by the Educational Research Center at Syracuse, New York: M. Marien, *Alternative Futures for Learning: An Annotated Bibliography* (Syracuse, 1971). A condensed version by the same author in which he confines himself to essential references on this subject still reaches two hundred titles.

Whereas in the early stages discussions and ideas were centred on what shape the education of tomorrow should take, the question 'for whom?' has now moved to the fore. T. Green, 'The Dismal Future of Equal Educational Opportunities' in *Futures* (Guildford, Surrey), 1972, takes the pessimistic view that the dream of equal opportunities in education will never become a reality, mainly for political reasons. A different view is represented in two reports, the first, which involved many years of investigations, under the auspices of UNESCO, the second on the initiative of the European Cultural Foundation as part of their 'Project 2000'. These express the view that equality of opportunity is not only a necessity, but also an educational objective which is capable of being attained: E. Faure (ed.), *Learning to Be* (UNESCO, Paris, 1972), and B. Schwartz, *L'éducation demain* (Education tomorrow: Paris, 1973).

These ideas have recently been under attack by American and British researchers as being illusory:

◇ Jensen, Herrnstein and Eysenck maintain, as a result of their research, that educational success, or the lack of it, is in the main determined by heredity.

◇ Coleman takes the view, following one of the most comprehensive social research studies ever undertaken – 600,000 pupils, 60,000 teachers and 4,000 schools took part – that it is not education in schools but the family situation in the home which plays the decisive part in children's development.

◊   Jencks, Gintis and Bowles have tried to show that class structures in society cannot be counterbalanced by equal opportunity in education. Schools, they say, are 'marginal institutions', and if we wish to transcend this traditional state of affairs we must introduce political controls over the economic structures which determine our society.

A new lever for change

Educationalists interested in futurology do not accept these obstacles as decisive. Jane Gaughan, 'Futuristics as a Subversive Activity' in *Trend* (Amherst, Mass.), Spring 1971, sees education for the future as a means of making the idea of a changing society acceptable as a matter of course. A detailed report by H. G. Shane, *The Educational Significance of the Future* (Indiana University, 1972), undertaken at the request of the government, shows the extent to which preoccupation with a different future and dissatisfaction with the present has taken hold in American schools and universities. The report contains an appendix with a bibliography of the latest publications on the subject of 'education and the future'.

Here is a short list of important publications on this subject:

M. Abrams, 'Mass Views of the Future' in *Futures* (Guildford, Surrey), June 1971. The author reports on a poll conducted by the Dutch Foundation for Statistics in which people were asked what change they would most like to see in the future and what they expected to happen. Ninety-one per cent gave as their first priority a lessening of class distinction; second place, with 90 per cent, fell to a desire for more working-class students at the universities.

H. v. Foerster, 'Perception of the Future and the Future of Perception' in *Instructional Science* (Amsterdam), March 1972, comments on the 'blindness' of those who do not wish to see.

B. Hayward, 'The Human Development Goal: Psychic Under-development and the Future of Education' in *Futures*, 1972, develops the basic idea that until now equality of opportunity has been used to give 'more people the chance of forging social weapons' with the help of which they then tried to improve their position in the social and economic hierarchies. Education should, in future, play a leading role in the effort for more co-operation and in reducing aggression. In place of the present-day climate of spiritual underdevelopment, a higher degree of spiritual development should be aimed at, attainable by nearly all people wherever they were born.

Alvin Toffler, 'The Psychology of the Future' in *Learning for Tomorrow: the role of the future in education* (New York, 1974), suggests that a combination of 'learning and action' could bring about social change.

J. Coleman and a commission of experts reached a similar conclusion in a study about the future of the education of minorities; they found that today's youth is overfed with information and hungers for experience, and should be given more opportunity to alternate study and work. Coleman suggests 'work communities' to include all age groups. In such a community of a thousand people there should be 90 children under the age of four, 180 between the ages of five and thirteen, and 100 people over sixty-five. (Source: *Time* (New York), 27 August 1973.)

Reports and case histories on the progress of future-oriented education can be found in *The Futurist* (Washington), August 1974.

A treatise on the sociocultural, biological and psychological environment for youth at the end of this century, based on an international meeting convened in September 1975 by the 'Jeugendprofil 2000 Foundation' in Amsterdam, has been published under the title *Adolescence and Youth in the Year 2000*, ed. J. P. Hill and F. J. Mönks (London, 1976).

The Revolt against Spurious Achievement

Uncareers

If one takes a closer look at who rebels against the 'achievement-oriented society' and why this rebellion is taking place, it will be seen that the protest is not directed against achievement itself, but against the wrong kind of achievement. In many cases, as I have been told in conversations with dropouts who were unable to stand the treadmill, it was precisely because they wanted to achieve something that they had to drop out. Programmed, mechanized and tyrannical methods of work demanded something of them which they were not prepared to give: slavery.

A young English couple from Birmingham, both scientists, having passed their examinations, prepared an address list for all those who wanted to have no more to do with dull, unintelligent drudgery: Ann and Martin Link, *Directory of Alternative Work* (298b Pershore Road, Birmingham 5). Their publishing organization bore the significant name 'Uncareers'. The directory appeared periodically. Only that work was listed which is interesting and which allows those doing it the freedom to organize their own time. Jobs offered included social work with old people, the mentally ill and alcoholics, posts in free schools or teaching gipsies, help with house-building, work in television studios, holiday jobs, and so on.

More and more people, even those in 'good jobs', would rather give up some of their earnings in exchange for more time, fresh air and work which is enjoyable. In 1972 a series of articles was published in the *Wall Street Journal* (New York) about such 'job deserters'. Pierre Drouin reports on the same phenomenon in France in *Le Monde* (Paris), 10 February 1973:

> Full time occupation of men in production jobs of forty to forty-five hours per week . . . is not really necessary. The individual can enrich his life intellectually by foregoing some paid work and

develop his creative faculties and ability to co-operate. The new 'logic of needs' . . . should guide us to the building of a system in which those wishing to do so can lower their living standards by doing without a number of products of the consumer society (which they regard in any case as serving artificial demands) and in return work only twenty hours a week. Part-time work is predicted for women. Why not also for men?

Young French people are already practising this, according to a study produced for the official Centre for Employment Studies in Paris: Jean Rousselet, *Bulletin d'information* (Paris, November 1972). Many young French people enter the normal, routine work process as late as possible. They often spend years in marginal jobs, as film extras, medical ancillary workers, opinion poll researchers, car-wash attendants, interpreters, tourist guides, jewellery-makers. Following a programme on French television about a girl who had left her home town and taken up work as a shepherdess in the country, the Ministry of Agriculture received around seventy thousand letters from young people wanting to do the same and asking how to set about it. *Les Métiers de la nature* (The natural professions: Paris, 1972), lists 233 jobs 'away from the towns' and has become a best-seller.

At the beginning of the sixties a report in the Catholic monthly magazine *Esprit* (Paris) told of the spiritual awakening and creative development in the villages of the 'Club Méditerranée' of people normally confined and restricted through their jobs. This is confirmed in a book by Christiane Peyre and Yves Renouard, *Histoire et légendes du Club Méditerranée* (Paris, 1971). Leisure time in these holiday colonies is often spent not passively but in adventures (trips to lonely islands in large sailing boats), discovery (deep-sea diving), artistic work (painting, sculpture), acting, building, cooking, composing, singing, discussions with stimulating people such as the sociologist Edgar Morin. The wearing of swimming clothes is an additional aid to democratization. The rapid expansion of the Club has, however, not been entirely beneficial to its *esprit*.

New higher needs

It is important to observe and record the phenomena of rebellion and the flight from boring and therefore tiring jobs as symptoms of the dissatisfaction with work which does not stimulate human faculties. In this area, too, criticism is beginning to bear fruit. Around the year 2000 the work situation will probably become one of the most intensive areas for new development and experiment.

An American psychologist who was an exponent of the so-called third force in psychiatric healing (after Freud and Jung), Abraham Maslow, predicted this development in *Motivation and Personality* (New York, 1960). In this book, as in many of his other works, the author develops his concept of a 'ladder of human needs'. Man is a being whose desires always lead one step beyond the one he has attained. As soon as he has achieved one of his desires, he will turn to the one above. At the lowest level are the requirements of the body: food, physical safety, shelter. Then follow his social needs: recognition by his fellow-men, a feeling of togetherness and belonging. Finally – and this is the stage which has now been reached in some of the developed countries – the ego demands self-expression and self-realization, full development of one's own faculties, creative possibilities. This explains the apparent paradox discussed by Martin Glaberman in his article 'The Rise of Militancy in the Auto Industry' in *Social Science and Modern Society* (New Brunswick, N.J.), 12 November 1972, whereby the car worker who can hardly wait to get out of the factory in the evenings spends his weekends working on his own car and considers this a satisfying occupation.

This and other articles which have appeared since the end of 1972 on the subject of rebellion in US production firms were occasioned by the continuous acts of sabotage in the most modern car factory in America, the largely automated Lordstown works. There is a report on this by Emma Rothschild, 'Automation and the workers at General Motors', in *The New York Review of Books*, 23 March 1972.

Between 1960 and 1969 the number of complaints at General Motors factories more than doubled, rising from 106,000 to 256,000 per annum, and has since risen, according to the latest estimates, by another 100 per cent. To that must be added the problem of absenteeism, not confined to the United States but noticeable also in Italy, France and the Soviet Union. According to statements by the director-general of Fiat, 14,000 out of a total of 91,000 employees stay at home each day.

There is a high percentage of younger, better trained workers who are more self-assured and less 'obedient'. This is developed by A. A. McLean and C. R. De Carlo, 'The Changing Concept of Work' in the management journal *Innovation* (New York) No. 30, 1972. It is shown that younger employees in particular demand not only higher wages and improved working conditions, but also a 'meaning' to their work. It is vital that management should recognize this desire for self-determination and decision-making and the longing to do that which one wants to do. If they fail to do so, they must expect growing difficulties during the coming years.

The French sociologist Dr A. Laville of the National Conservatory of the Arts and Professions in Paris comments on the bad working conditions which still prevail in the majority of French factories: 'We know that they can lead to difficulties and, as a result of nervous exhaustion, to actual change in personality' (quoted in an article by Daniel Garric, 'Le travail contesté' (Work contested) in *Le Point* (Paris), 30 April 1973).

Experiments in the work sphere

The following are the results of a number of research projects over the last few years:

US Health Education and Welfare Department, *Work in America* (Washington, 1971);

S. Marcson (ed.), *Automation, Alienation and Anomie* (New York, 1970);

M. Argyle, *The Social Psychology of Work* (London, 1972);

L. E. Davis and J. C. Taylor, *Design of Jobs* (London, 1972): an excellent reader containing twenty-nine articles ranging from historical experiments to thoughts on the future;

F. Herzberg, *Work and the Motivation of Work* (New York, 1970) and *The Motivation to Work* (New York, 1959): reports on a variety of experiments inside the existing economic structure; Herzberg is the advocate of job enrichment which aims to take monotony out of routine work by introducing more variety and self-determination in planning one's own work;

E. Jacques, *Work, Creativity and Social Justice* (London, 1964): the author is a psychologist who co-operated with the progressive director of the Glacier works near London on the so-called 'Glacier Experiment', which offers an opportunity for the development of initiative and creativity to its workforce;

R. Poor (ed.), *Four Days, Forty Hours* (London, 1972): develops the idea of a four-day week;

D. Jenkins, 'Democracy in the Factory' in *The Atlantic Monthly* (Boston), April 1973: the interesting experiment described in this article was carried out by the giant cosmetics manufacturers Procter & Gamble, well known for their aggressive management policies.

However, all these experiments have their drawbacks. They are only kept going if they prove profitable. When a crisis occurs they are discontinued. This is dealt with in the article 'Where being nice to workers didn't work' in *Business Week* (New York), 20 January 1973.

The electronics company Non Linear Systems in San Diego tried nearly all the prescriptions for modern work psychology. Productivity rose by 30 per cent, sales and profits increased. The experiment was mentioned in several newspaper articles as a pioneering venture. Then came the crisis in the aerospace industry. Receipts fell by one half. The owner lost his nerve. Everything that had previously been tried out and publicly praised was now wrong – too little control, too much talk, over-high wages – and the workers were said not to be equal to the 'new style'. 'Some of them need routine work,' the boss now said in an equally convincing tone.

Dignity at work

The new emancipated style of work must not be guided by profit motives if it is to succeed, but must rest on a social and ethical basis. These considerations prompted Ernest Bader, a Swiss who emigrated to England in 1920, to try out the following successful experiments at his chemical factory in Woolaston. First he introduced profit-sharing for his employees. Then, in two stages, in 1951 and 1963, he made all his employees co-owners with equal rights in a 'commonwealth'. Not only are the conditions at the works humane, but the employees have imposed humane controls on themselves with regard to one another and their environment. The factory will be limited to a workforce of 350, even if business goes well, in order to retain its human scale. (Sources: *Current Journal of Social Issues* (Lakeside), Winter 1971; E. F. Schumacher, *Small is Beautiful* (London, 1973); F. H. Blum, *Work and Community* (London, 1968).)

Common ownership is practised in France in the factories of Marcel Barbu, and in West Germany in the cases of the Behrens company at Ahrensburg, Photo-Porst in Nuremberg and the co-operative steelworks at Süssmuth about which a detailed report was published by Franz Fabian, *Arbeiter übernehmen ihren Betrieb* (Workers take over their factory: Hamburg, 1972).

In March 1972 the 'Carl-Backhaus-Stiftung für Demokratie in der Wirtschaft' (Carl Backhaus Foundation for Democracy in the Economy), Ahrensburg, West Germany, held a conference under the direction of F. Vilmar which was reported in the book *Menschenwürde im Betrieb* (Dignity at work: Hamburg, 1973); it set out to discuss the problems and the already existing models of humane workplaces. The objectives of these were described to participants and observers at this conference as follows:

> In this country there is much talk about making man the focal point at his place of work. In practice there is not much evidence of this. And yet it is not the much-cited 'objective necessity' which produces inescapably authoritarian and monotonous working conditions. There exist remarkable blueprints – sometimes already being successfully put into practice – for shaping conditions under which production is organized in a more democratic and humane fashion, even under private ownership. This proves that the inhumane atmosphere of alienation and monotony which still dominates the world of industry, as well as that of office work, is by no means unalterable and can be abolished. The aim is abolition of all exploitation of man, by private owners or the state.

(Sources: F. Vilmar (ed.), *Menschenwürde im Betrieb* and *Demokratisierung* (Democratization: 2 vols., Neuwied, 1973).)

Criticism of Science and Technology

Before and after Hiroshima

Earlier criticism of technology, which was in the main conservative and emanated principally from Europe – Georges Bernanos, Georges Duhamel, Friedrich Georg Jünger – broadened from 1945 onwards, under the impact of the rapid growth of technological power, to a general criticism of science (as the basis of technological inventions) and of society (held ultimately responsible for the setting of scientific objectives). The principal impetus to look at science and its social conditioning through Marxist eyes came from J. D. Bernal, *The Social Function of Science* (London, 1939) – a monumental work, the powerful influence of which only began to make itself felt after the Second World War. Determined self-criticism and a pointer to possible dangers, however, can already be found in Robert K. Merton, *Science and the Social Order* (1938) and *Science and Democratic Social Structure* (1942).

In the shadow of Hiroshima and Nagasaki, science's questioning of its bad conscience becomes a kind of scientific discipline in its own right, though pursued from various standpoints and in differing directions:

◇ criticism of the aims and dependence of technology;
◇ criticism of technology's dogmatism and thirst for power.

Criticism of aims and dependence

This is dominated by the direct or indirect dependence of technology on the armaments industry, which increased enormously during the Second World War and the period following it.

Periodicals
In 1945 the editor of the *Bulletin of Atomic Scientists* (Chicago), Eugene Rabinowitch, initiated a debate about these problems from a

liberal point of view. This periodical continues today under the title *Science and Public Affairs*, and the subject-matter has been broadened to include all public affairs connected with science.

Since 1968, a publication has appeared every other month in the US which is far more critical and radical than the *Bulletin: Science for the People* (9 Walden Street, Jamaica Plain, Massachusetts), the organ of SESPA (Scientists and Engineers for Social and Political Action), which is mainly aimed at activists of the Left.

A number of scientific journals in the US deal with specialized branches of science and technology:

The Insurgent Sociologist (University of Oregon);

Rough Times, formerly *Radical Therapist* (W. Somerville, Mass.);

Physics Free Press (Boston);

Interrupt – Computer People for Peace (Brooklyn, New York).

In addition, there are frequent special publications which deal with topical subjects: in 1972, for example, students and faculty members at Harvard University produced *The University–Military Complex*, designed to highlight the part played by university institutions in the research on armaments projects.

In Britain *Science for the People* (9 Poland Street, London, W.1), the organ of the British Society for Social Responsibility in Science, is published every other month.

The London weekly *New Scientist*, which sells a fairly large number of copies and reaches a standard of topical, serious and committed scientific journalism unequalled anywhere in the world, has become increasingly critical of social and scientific matters since the beginning of the seventies.

In July 1970 the periodical *Survivre* (Survival: 1 rue de Prony, Paris 19) was started by C. Chevally and A. Groethendieck with the following declared objective: 'A group of scientists has come to the conclusion that the common fight for the survival of scientists and non-scientists of all nations must become identical with the aim of a renewal of life.'

Books

A broadly based survey of the self-analysis of science is provided by Jerome R. Ravetz, *Science and its Social Problems* (London, 1971).

The dependence of science and technology on industry and government are forcibly described in:

H. L. Nieburg, *In the Name of Science* (Chicago, 1966);

D. S. Greenberg, *The Politics of Pure Science* (New York, 1967);

Seymour Melman (ed.), *The War Economy in the United States* (New York, 1971).

In Britain and the US a considerable number of books have been devoted to research on armaments and the technology of war. Here is a selection:

Robin Clarke, *We All Fall Down* (London, 1968): deals with biological and chemical warfare;

—— *The Science of War and Peace* (London, 1971);

Nigel Calder (ed.), *Unless Peace Comes* (London, 1968): a series of articles about future arms developments;

F. Lapp, *Arms beyond Doubt* (New York, 1971): about the tyranny of the armaments industry.

A more traditional viewpoint which welcomes technology as a decisive element in modern warfare is represented by Stephen T. Possony, *The Strategy of Technology* (Cambridge, Massachusetts, 1971).

J. Rotblat, *Pugwash – the First Ten Years* (London, 1967), is a somewhat dry but complete report about East–West conferences on peace and disarmament held by scientists.

Finally Seymour Melman, in *Our Depleted Society* (New York, 1965), *Pentagon Capitalism* (New York, 1970) and *The Permanent War Economy* (New York, 1974) destroys the myth that defence spending has been good for the American economy. Far from bringing prosperity, he contends, the armament race has lowered US industrial efficiency and triggered the inflation of the 70s.

Articles

Paul Doty, 'The Community of Science and the Search for Peace' in *Science* (Washington), 10 September 1971, is of interest here.

An important article which the author intends to enlarge into a book and which opposes technical innovation as the motivating force of a society unchanged in its basic inhumanity is J. McDermott, 'Technology – the Opium of the Intellectuals' in *New York Review of Books*, July 1969. It sharply attacks a technological intelligentsia who have accepted the principle of *laissez-innover* as a modernized version of the *laissez-faire* doctrine.

J. McCaull, 'The Politics of Technology' in *Environment* (St Louis, Mississippi), March 1970, is a polemic against the subjection of science and technology to the aims of increased productivity and the utilitarian principle that material gain equals human well-being.

In 'A Hippocratic Oath for Applied Scientists' in the journal *New Scientist* (London), 7 January 1971, M. W. Thring suggests an oath for scientists comparable to the hippocratic oath, the unwritten law of the medical profession. He follows up his suggestion in the book *Man, Machines and Tomorrow* (London, 1973). The shortened form of this oath which he proposes is:

I vow to strive to work towards the co-existence of all human beings in peace and human dignity with all the necessities for a self-fulfilling life and freedom from fear, stress, ugliness, pollution and noise.

Criticism of dogmatism and power-seeking

A number of scientists and technologists are not content with criticizing objectives but, especially since the sixties, are trying to delve deeper and ask themselves whether the causes for the 'fall from grace' of science and technology are not to be found in their own method of thought, their dogmatic attitude with regard to other concepts, life-styles and values.

An eight-page periodical entitled *Manas*, small in size but extremely influential, published since the end of the Second World War in Los Angeles (P.O. Box 32112, El Sereno Station), voiced this criticism at a time when those concerned were still looking for the reasons for their dilemma outside science, among the tempters and seducers of 'pure science' and 'sound technology'.

Amongst books, a pioneering role must be assigned to the works of Lewis Mumford, the American historian and cultural philosopher: Lewis Mumford, *Technics and Civilization* (New York, 1934); —— *The Condition of Man* (New York, 1944); —— *The Myth of the Machine* (New York, 1966); —— *The Pentagon of Power* (New York, 1970). Mumford depicts the relationship of man and his tools which, finally, by their mechanization not only enslave him but destroy the world. His hope is centred on 'biotechnics', a radically new model derived not from machines but directly from living organisms and organic complexes (ecosystems), which would eventually replace the 'feeding' of machines with human work by the fuller development of human potential.

No such hopeful outlook is offered by the French Protestant theologian Jacques Ellul, whose book *La technique, ou l'enjeu du siècle* (Paris, 1954) remained practically unnoticed until it was discovered and translated by the American historian and philosopher John Wilkinson under the title *The Technological Society* (New

York, 1964), when it became a much discussed standard work. Ellul describes how the tyranny of technology, which is not only a physical but also an intellectual force, increasingly encroaches on life, humanity and liberty. When finally instinct and mind have been integrated – what remains?

In the sixties, young people tended to question the values underlying modern science and technology only spontaneously and without deeper analysis. Theodore Roszak is an exception. In two works, *The Making of a Counterculture* (New York, 1968) and *Where the Wasteland Ends* (New York, 1971), he identifies the 'myth of objectivity' as the root of the scientific approach and technocratic domination. In an essay entitled 'The Autopsy of Science' in *New Scientist* (London), 11 March 1971, he comes to the following conclusion with regard to this changed attitude in young people:

> We are at a moment when the reality to which scientists address themselves comes more and more to be recognized as but one segment of a far broader spectrum. Science, for so long regarded as our single, valid picture of the world, now emerges as, also, a *school*: a school of consciousness, beside which alternative realities take their place.

The New Science Group which was formed at the beginning of the seventies at the University of Michigan by John Platt, Don Michael, Peter Harper, Robert Olson and Mark Berg attempts to see the 'crisis in science' as a preparatory period for an 'extended science' which would have certain basic characteristics in common with 'old science' but would differ from it in other ways. The group lists the following points as reasons for its doubts regarding the role of science and technology:

◇ the failure of technology to solve certain problems, its unforeseen side-effects and the inability to predict such effects at all levels;

◇ the fact that certain problems can be solved more easily if an apparently irrational attitude is adopted (the cybernetics expert Gregory Bateson has pointed out that the biological processes on a lower level work very well if left to themselves: good gardeners – and doctors! – often achieve better results in this manner);

◇ the suspicion that there may be sociological, psychological and logical limitations to scientific knowledge.

Assessment and Control of Technology

The Office of Technology Assessment

On 13 October 1972, a law was passed by both houses of the US Congress setting up a parliamentary office for a forward-looking assessment of planned technological innovations. This Office of Technology Assessment (OTA) was given a budget of $5 million annually for the fiscal years 1973 and 1974. It has a staff of between forty and sixty experts. The principal tasks of the new office are outlined as follows:

◇ assessment of actual and probable effects of technology;
◇ suggestions for alternative technological methods for carrying out certain projects;
◇ estimation and comparison of the effects of different methods and programmes;
◇ submission of the results to the relevant legislative bodies.

The work of this body might be criticized on the following grounds:

◇ Publication of the results and the information on which they are based is not legally required. It depends in each case on the agreement of the Technology Assessment Board, a directorate of thirteen members – six Senators, six members of the House of Representatives and the Director of the OTA.

◇ The new office has only consultative and no controlling functions.

◇ The studies are undertaken by outside bodies. There is a possibility that the experts who are consulted may frequently be the same persons who have a special interest in the success of a new technological development. Experience hitherto with special committees which have had to investigate the problems of aviation or the automobile, as well as with the Congressional Research Service, has frequently shown an uncritical dependence on experts who have in many instances also been working for industry.

◇ Since the OTA could only be set up in the face of great difficulties and strong opposition it is likely, for the time being, to avoid the really 'hot' issues, such as the priority of public transport over the private car, and turn towards future technological possibilities which have not yet been disputed, such as biological and genetic intervention or the consequences of peaceful uses of atomic fusion.

International growth of TA

The growing interest in TA is shown by the fact that during 1973 three international conferences, at Gargano in Italy, Salzburg and The Hague, discussed the new subject. The Council of Europe was considering the formation of a European agency for anticipatory assessment of technology. In Holland, France, Sweden, the USSR, Canada and Japan the creation of such bodies is in preparation.

In West Germany, Dr Weyand of the parliamentary Scientific Services in January 1973 prepared a memorandum regarding the 'necessity, tasks and work methods of a department at the Federal Assembly for the systematic analytical assessment of scientific and technological developments'. His reasons, among others, were:

◇ The traffic problem in West Germany threatens to become a 'national catastrophe'.

◇ There are signs that the government's subsidies to the various scientific and technological development areas do not correspond to the actual needs of society in the country.

◇ Whereas in the fields of science and technology an increasing number of officials, staff and experts are employed on the executive side, expansion of controls by the legislature has been neglected. (Source: Doc. Reg. No. WF VI – 23/73.)

Public participation

Tay Wilson of Nipissing College, Ontario, Canada, on his own initiative undertook an opinion poll of the population of this small town on the probable consequences of a withdrawal of all public transport. The result caused the government to refrain from taking this step. (Source: Minutes of the International Society for Technology Assessment at The Hague, May 1973.)

Anthony Wedgwood Benn, former British Minister of Technology, pleaded in the House of Commons as Opposition Spokesman on Trade and Industry for more participation by ordinary people in taking technological decisions (*New Scientist*, London, 24 May 1973):

The problems of technology assessment cannot be resolved by stuffing computers with specially commissioned economic, social and psychological data. Even if every single factor could be fed in, and properly weighted – which is impossible – people would not accept the resultant decision, simply because they had played no direct part in reaching it. When we talk about participation, or 'assessment done in the light of a wide range of studies', we are talking management language with all the dangers of human manipulation that this implies. This sort of 'participation' is no substitute for democratic control. . . .

There is no predetermined future that we have to accept, nor is there any specialist entitled to claim a monopoly of wisdom, in telling us what it is. Nor is it true that, as scientific and technical skill increasingly reveals the laws of nature, and helps us to use them, our freedom and happiness will automatically expand in proportion to our knowledge or our material power. The truth may prove to be the exact opposite. As technical power increases, mankind's apparent conquest of nature may produce new tyrannical organisations to organise that 'conquest', and they then extend their domain over their fellow men. . . .

Many technological decisions are virtually irreversible once they have been reached, and until we strip away unjustifiable secrecy we can have no real democracy.

Subjects already investigated

There have been investigations into:
◇ the siting of atomic power stations (Sweden);
◇ the siting of a third London airport (Britain);
◇ the construction of tall buildings in a country threatened by earthquakes (Japan);
◇ the consequences of cybernetic teaching (Japan);
◇ artificial human organs (Japan);
◇ the control of car exhaust gases; computer and communications networks; the industrial use of enzymes; marine cultivation (ocean farming); water pollution (MITRE Corporation, the think tank of the US government);
◇ the consequences of the use of antibiotics in agriculture (Britain);
◇ the effects of cable television (US);
◇ new food sources such as seaweed, etc. (US).

Literature on TA

An indispensable item for the study of events leading up to the inception of TA is the handbook which furnished the arguments for the fight against supersonic aircraft: W. A. Shurcliff, *S/S/T and Sonic Boom Handbook* (New York, 1970).

Marvin J. Cetron (ed.), *Technology Assessment in a Dynamic Environment* (New York, 1972), contains forty-two articles by international experts.

The periodical *Technology Assessment* has been published since 1972 by Gordon and Breach (New York and London). Volume 1, No. 1, pp. 75–83 contains a detailed bibliography.

New Scientist (London), 24 May 1973, contains several articles which provide a good survey of the problems of TA, to which this number is entirely devoted.

Professor M. S. Baram of MIT, in an essay entitled 'Technology Assessment and Social Control' in *Science* (Washington), 4 May 1973, looks critically at the difficulties encountered by TA through secrecy in the private sector.

Valuable information about the history and prospects of the OTA is contained in an article by Anne H. Cahn of MIT and Joel Primack of Harvard in *Technology Review* (Cambridge, Mass.), March/April 1973.

Harold Green, writing in the periodical *Business and Society Review* (Boston), Spring 1973, expresses serious concern about TA in 'Technology Assessment and Democracy'. His main argument is that, due to TA, the advantages of a new technological development tend to be overestimated and its risks underestimated in its early stages. The effects could therefore be exactly opposite to what was intended: potentially dangerous developments receive an official stamp of approval through a preliminary examination by experts.

Also of interest are:

Vary T. Coates and John E. Mock, *Summary of the Southern Regional Conference on Technology Assessment* (National Science Foundation, Washington, May 1974);

Joseph F. Coates, 'Some Methods and Techniques for Comprehensive Impact Assessment' in *Technological Forecasting and Social Change* (New York), Vol. 6, No. 4, 1974.

Possibly the most important research on this subject is that being carried out by the Technology Assessment Group set up within the framework of the Program of Policy Studies in Science and Technology at George Washington University, Washington D.C.

New Technologies

Soft technology

The spiritual fathers of 'soft technology' are the Russian Kropotkin and the American Thoreau. One of its pioneers who is still alive today is Peter van Dresser, born in Holland. As early as 1938 he wrote an essay, now frequently quoted, in *Harpers* magazine (New York), with the title 'New Tools for Democracy'. In it he outlined how research and technology can help to remove the excesses of a technology which threatens man and his environment and prepare a 'non-technocratic future on a human scale'.

It is evident that the concept of an 'alternative technology' is gradually being taken seriously. The magazine *Impact*, published by UNESCO in Paris in English and French, devoted a special number to this subject (October–December 1973). Contributors included Robin Clarke (UK), Peter Harper (UK), Mansur Hoda (India), Josefina M. Abraham (Mexico), Philippe Arrêteau (France) and Robert Jungk (Austria).

The following periodicals are especially concerned with soft technology and other forms of technology favourable to the environment:

A.S.E. – Alternative Sources of Energy (Minong, Wisconsin, Box 36B): every number expressly points out that '*A.S.E.* is *not* protected by copyright';

Mother Earth News (Madison, Ohio);

The New Alchemy Institute Bulletin (Woods Hole, Mass.);

Resurgence (Kingston-on-Thames, Surrey);

Undercurrents (London, N.W.6): this paper regularly publishes bulletins of addresses of soft technologists all over the world.

Articles on this subject appear frequently in publications concerned with the environment, such as:

Architectural Design (London, W.1);

The Ecologist (Richmond, Surrey);

The Environment (St Louis, Mississippi);
Le Sauvage (Paris);
Umwelt (Environment: Düsseldorf).

An interesting discussion in the form of two articles, one for and one against soft technology, was published in *New Scientist* (London), 11 January 1973.

The most thorough and self-critical study in depth of soft technology, its problems and prospects so far undertaken was produced by Peter Harper of Hove, Sussex, for a conference of the UNESCO youth organization under the title 'Soft Technology. A Proposal for Alternatives under Conditions of Crisis' (Paris, 1972) – a mimeographed manuscript. He continued this self-critical approach in Nos. 5 and 6, 1973, of *Undercurrents*.

An excellent politically oriented book on new technologies is David Dickson, *Alternative Technology and the Politics of Technical Change* (London, 1974).

Various pamphlets have been published by groups working on soft technology in California, New Mexico, Costa Rica and Nova Scotia.

Special mention should be made of detailed 'experience reports' by John Todd of Cape Cod, Massachusetts, and by the engineer B. Eriksson of Bybaken 1, Värmdö, Sweden.

An outline worth reading for plans to form a 'biotechnical community' which have since been put into practice by Robin Clarke, his wife Janine and their friends was published in *Futures* (Guildford, Surrey), June 1972.

It is worth noting that in recent times, presumably under the impact of the environmental crisis, the East European countries, too, have begun to develop a critical appreciation of technology, a process which had started in Czechoslovakia in 1968 and was then interrupted. A member of the Polish Academy of Science, the sociologist A. Sicinski, stated in a paper presented to an international congress for research into the future in Rome at the end of September 1973: 'It may be too optimistic to say that man will *have* to change his attitude towards technology. But it can justifiably be stated that the necessity for this is being felt and that certain changes in that direction are already discernible.' He quotes with approval A. Kepinski, a psychiatrist living in Cracow, who makes the following statement in his book *Rytm zycia* (Rhythm of life: Cracow, 1972):

Man conquers the world. Often machines seem more important than man, the sole measure of his achievement. The environment is turned into an emotional desert, or even an enemy, which can be dealt with at will, according to momentary requirements. Since the world of man is, above all, a social world, it follows that this attitude will be applied to individuals and whole communities. Man becomes a more or less efficient cog in the machinery, a part which needs to be overhauled or renewed from time to time. Society is seen as a complex mechanism of millions of wheels and gears (the individuals) which can be arranged, driven or turned off without limitations. It is evident that such a concept of the human world – and indeed of nature – simply does not correspond to the truth.

Conversations with people in the eastern European countries lead me to hope that in the Soviet Union and in Hungary too efforts are being made to further an alternative technology. There is an indication of this in the above-mentioned book by the Polish psychiatrist:

> It can be said that, until recently, technology stood in active opposition to nature. It tried to subdue the forces of nature. Thanks to technology, man has acquired mastery over his natural environment and created an artificial sphere of life in which he does not feel at ease. He has now reached the point where he must cease to dominate nature and is faced with the need to change the attitude of technology towards nature from an antagonistic one to a symbiotic one. A symbiosis of technology and nature will make man's environment less artificial and monotonous and lend it an individual touch. The first signs of such co-operation are already apparent. We have at last begun to mould technology according to the pattern of nature . . .

Alternative technology in developing countries

In recent years, numerous reports from Western travellers in China point to a different style of industrialization – decentralization, home production of tools and machinery, etc., as reported by G. Goodman in *New Statesman* (London), 15 June 1973. At the conference on the environment in Stockholm, the Western environmental group 'Dai Dong' (which is closely connected with the World Council for Reconciliation) developed a close contact with the Chinese delegation and arranged a meeting with their speakers. The *Peking Review*, which is obtainable in the West but which, due to its propagandistic double-talk, makes very heavy reading, also contains various pointers in that direction.

Professor Ernst Florian Winter of Schloss Eichbüchl near Wiener Neustadt made a detailed study of the environmental question during a prolonged stay in China, and is writing a book about the 'Chinese road'. L. A. Orlans and R. P. Stuttmeier published an article in *Science* (Washington), 11 December 1970, entitled 'The Mao Ethic and Environmental Quality', which is still the best source with regard to this problem published in the West.

An informative talk by Derek Bryan, 'Lessons We Can Learn from China', is printed in the volume M. Schwab (ed.), *Teach-in for Survival* (London, 1973). The monumental historical study by Joseph Needham, *Science and Civilization in China* (Cambridge, 1952–70), shows that the wisdom with which Chinese communism has tackled its technological development has its roots in historical traditions.

Ignacy Sachs, the Polish economist living in Paris, and a study group for developing countries formed in Starnberg at the 'Max-Planck-Institut für die Erforschung der Lebensbedingungen in der wissenschaftlich-technischen Umwelt' (Max Planck Institute for research into living conditions in the scientific and technological environment), are attempting to find new solutions which correspond to the real needs of the Third World.

The negative effects on the developing countries of blind scientific and technological 'progress' are described by Giovanni Rossi in his essay 'La Science des Pauvres' (The science of the poor) in *Recherche* (Paris), 30 January 1973:

> The systematic application of science and technology to manufacturing processes has resulted in a considerable rise in productivity in capitalist countries, whereas the situation in the developing countries has remained stationary. The increase in manufacture of synthetic materials alone (synthetic rubber manufacture from 0·6 per cent in 1938 to 56 per cent in 1965) has caused unemployment for large numbers of people in countries providing raw materials.

So does the answer lie in the industrialization of the Third World? High patent fees would have to be paid to the West. Furthermore, the transfer of technology would create a group of captives. The new Western industries established in the Third World would be financially and technically dependent (spare parts!) on the West. This explains the strong attack conducted by René Dumont in *L'utopie ou la mort* (Utopia or death: Paris, 1973) against this form of conquest dressed up as aid.

A solution may be found, beside soft technology, in the form of adapted technology which reduces unemployment and encourages independence. The founder of the Intermediate Technology Development Group, E. F. Schumacher, writes in his book *Small is Beautiful* (London, 1973):

> As Gandhi said, the poor of the world cannot be helped by mass production, only by production by the masses. The system of *mass production*, based on sophisticated, highly capital-intensive, high energy-input dependent, and human labour-saving technology, presupposes that you are already rich, for a great deal of capital investment is needed to establish one single workplace. The system of *production by the masses* mobilises the priceless resources which are possessed by all human beings, their clever brains and skilful hands, and *supports them with first-class tools*. The technology of *mass production* is inherently violent, ecologically damaging, self-defeating in terms of non-renewable resources, and stultifying for the human person. The technology of *production by the masses*, making use of the best of modern knowledge and experience, is conducive to decentralisation, compatible with the laws of ecology, gentle in its use of scarce resources, and designed to serve the human person instead of making him the servant of machines. I have named it *intermediate technology* to signify that it is vastly superior to the primitive technology of bygone ages but at the same time much simpler, cheaper, and freer than the supertechnology of the rich. One can also call it self-help technology, or democratic or people's technology – a technology to which everybody can gain admittance and which is not reserved to those already rich and powerful.

Either native developers receive training from the ITDG in the United Kingdom, or experts are sent out, not in order to form bridgeheads for Western industries, but to return as soon as possible. Schumacher describes their tasks:

> But we do know something about the organisation and systematisation of knowledge and experience; we do have facilities to do almost any job, provided only that we clearly understand what it is. If the job is, for instance, to assemble an effective guide to methods and materials for low-cost building in tropical countries, and, with the aid of such a guide, to train local builders in develop-

ing countries in the appropriate technologies and methodologies, there is no doubt that we can do this, or – to say the least – that we can immediately take the steps which will enable us to do this in two or three years' time. Similarly, if we clearly understand that one of the basic needs in many developing countries is water, and that millions of villagers would benefit enormously from the availability of systematic knowledge on low-cost, self-help methods of water-storage, protection, transport, and so on – if this is clearly understood and brought into focus, there is no doubt that we have the ability and resources to assemble, organise and communicate the required information.

The ITDG is aware of its own possibilities and limitations. Its founder comments:

> The applicability of intermediate technology is, of course, not universal. There are products which are themselves the typical outcome of highly sophisticated modern industry and cannot be produced except by such an industry. These products, at the same time, are not normally an urgent need of the poor. What the poor need most of all is simple things – building materials, clothing, household goods, agricultural implements – and a better return for their agricultural products. They also most urgently need in many places: trees, water, and crop storage facilities. Most agricultural populations would be helped immensely if they could themselves do the first stages of processing their products. All these are ideal fields for intermediate technology.

A similar approach is practised by John Morgan. His 'Village Technology Innovation Experiment' (Addis Ababa, P.O. Box 31) is intended to pass on useful knowledge. He has produced an 'Almanac for the Ethiopian peasant', started libraries with reference books in various villages and, together with the Abyssinians, developed prototypes of small technical appliances such as bread ovens, water heaters running on solar energy, water reservoirs and so on.

Additional bibliographical material on the subject of technology and the Third World:

I. Sachs, *La découverte du tiers monde* (The discovery of the Third World: Paris, 1971);

R. Dumont, *L'utopie ou la mort* (Utopia or death: Paris, 1973);

I. Illich, *Tools for Conviviality* (London, 1973);

A. Wedgwood Benn, 'China – Land of Struggle, Criticism and

Transformation' in *New Scientist* (London), 6 January 1972; Chi Wei, 'Turning the Harmful into the Beneficial' in *Peking Review*, 28 January 1972.

Evolutionary technology

Felix von Cube, writing with great clarity about cybernetics as the *Technik des Lebendigen* (Technology of the living: Stuttgart, 1970), a superb example of bridging the gap between the expert and the public, makes the following prediction:

> Without doubt man will be exploring more and more organic techniques in the near future and will attempt to imitate them by machines. We know today that nature possesses an enormous and many-sided potential of so far undiscovered techniques. Cybernetics has made giant strides towards the conquest of this potential. Further results reaching as far as artificial intelligence or even artificial life can surely be expected.

This type of research is of great importance since it attempts to bring about an evolutionary change in technology which could reduce the conflict between man and his tools.

The basis for this development was provided above all by Norbert Wiener, Warren McCulloch, Gordon Pask, Heinz von Foerster and Walter Rosenblith.

Among the basic literature on the subject we find the following:

G. Pask, *An Approach to Cybernetics* ((London, 1961);

G. Simondon, *Du monde d'existence des objets techniques* (The world of technical objects: Paris, 1958);

H. v. Lier, *Le nouvel age* (The new age: Liège, 1962);

W. Brodey, 'Biotopology 1972' in *Radical Software* (New York), Summer 1971.

Less well known are the pioneering essays published by Warren M. Brodey, together with the scientific journalist Nilo Lindgren, in the trade journal of the American electronics engineering business, *IEEE Forum* (New York): 'Human Enhancement through Evolutionary Technology' (September 1967) and 'Human Enhancement. Beyond the Machine Age' (February 1968).

Brodey's own experimental workshop, 'Electronic Tool and Toy' (Armory Road, Milford, New Hampshire), has had to be closed for the time being owing to lack of funds. But experiments are continuing. Most success has been achieved by architects experimenting with 'living', biological building methods:

◇ The work of the Architecture Machine Group at MIT is described by Nicholas Negroponte in several books, for example *The Architecture Machine* (Cambridge, Mass., 1969).

◇ The 'Institut für leichte Flächentragwerke' (Institute for light supporting structures) in Stuttgart is directed by Professor Frei Otto: see Frei Otto, 'Biologie und Bauen' (Biology and building) in the journal *Der deutsche Baumeister* (The German builder: Munich), February 1973, as well as various other publications of the Institute, which collaborates closely with Professor J. G. Helmcke's Institute of Biology and Anthropology at the Technical University in Berlin. Structural patterns of silicone algae invisible to the naked eye are stereoscopically photographed and used as models and patterns for the new principles of construction.

◇ Rudolf Doernach, also of Stuttgart, is working on new biological building materials designed according to the principles of the formation of micro-organisms.

◇ Ingo Rechberg and his team at the Technical University in Berlin are attempting to apply principles of biological evolution to technical developments: see I. Rechberg, *Evolutionsstrategie* (Evolution strategy: Stuttgart, 1973).

Another field in which 'living technology' is making progress is that of medicine. The following is a selection of papers throwing light on this:

'Designing better Heart Valves' in *New Scientist* (London), 11 January 1973: a report on successful attempts at the University of Utah to improve artificial heart valves;

'Enfin une vraie main artificielle!' (At last a real artificial hand!) in *Science et Vie* (Science and life: Paris), February 1973: a report on successful attempts by Schmidl, Zarotti and Zagnoni at the Institute of Vigorso di Budrio in Italy to amplify man's weak muscular electricity to such an extent that it can power a mechanical hand with thousands of component parts;

'Visionary on a Golden String' in *Fortune* (New York), June 1973: a report about successful attempts by the development firm of Alza in Palo Alto, California, to incorporate miniature auxiliary instruments into the human body in order to compensate for physical deficiencies or to introduce drugs directly into the affected organs by by-passing the bloodstream and the digestive tract.

There are numerous examples in a fascinating book by the scientific editor of the *Financial Times*: David Fishlock, *Man Modified* (London, 1969).

Progress in the fields of micro-electronics, membrane research and, above all, the study of how the individual organs of the highly complex precision-made human system and that of other living organisms co-operate and complement each other will exert an ever increasing influence on the technology of tomorrow.

Already in discussions about vital technical installations, the failure of which would prove fatal (for example power stations), it has been suggested that duplication on the lines of important organs in the human body (kidneys, lungs) might be necessary. The inoperative technical 'organ' could thus be instantly replaced, or at least partially replaced, by its duplicate.

A technology based on logotherapy as developed by Viktor Frankl of Vienna, which would make creative work possible, is described by E. Matchett in *Towards a New-World Technology* (Bristol, 1973). The firm of Matchett Training (14 Montrose Avenue, Bristol) advises firms on the introduction of FDM (Fundamental Design Method) and methods of logosynthesis.

Finally, two publications from among the growing number concerned with the design aspects of a new technology which seem to me especially interesting and productive are V. Papanek, *Design for the Real World* (New York, 1970), and a printed report about a conference held in 1971 in Manchester to discuss possibilities for democratic participation in technological, urban and other public projects: N. Cross (ed.), *Design Participation* (London, 1972).

Guided technology

In July 1967 the American economist Robert Heilbronner published an article in *Technology and Culture* (Detroit) with the significant title 'Do Machines Make History?' These few words sum up the tenor of the debate on the subject at that time. They start from the assumption that technology and its growing influence will determine or even dominate future development. Four years later the politically and economically rather conservative OECD published a study by an international commission about the 'social cost of fast economic growth'. This makes a determined plea for society to guide the development of technology. When considering which technical developments should be supported, the criteria of 'technical feasibility and commercial prospects' no longer suffice. Report and commentary in *Scientific American* (New York), August 1971.

Marxist critics maintain that only a socialist society can successfully tame technology. This point of view is forcefully presented in

David Dickson, *Alternative Technology and the Politics of Technical Change* (London, 1974).

It is considered possible in the West that technology can be guided towards 'beneficial goals' without drastic changes in the existing power structures of society. The belief that it is possible to shape the future – and that means also the future of technology – was expressed by the leading American futurologist Olaf Helmer – then ahead of his time – in his book *Social Technology* (New York, 1966).

The question of what the priorities should be now becomes the central problem, and it is significant that in 1972 a new periodical was devoted to this: *National Priorities* (New York).

The large number of wrong predictions about the future of technology point to hitherto neglected factors which would have to be brought into play in guiding technological development:

◇ the influence of an informed public prepared to take part in debating and deciding upon priorities;

◇ the possibility of keeping planning and development open and flexible so as not to impede the future through the over-rigidity of present planning;

◇ a far higher regard for the largely unpredictable 'human factor'.

New Democratic Institutions

Can the citizen take part?

The increasing number of civic initiatives in Western countries should be seen not merely as isolated incidents, but as an expression of real demand. The citizen has become tired of his habitual role, which allows him no say apart from his vote for a party or a candidate on whom he has little or no influence. He wants to participate actively. But where, how, and with whom?

Many towns have town halls but no citizens' halls. The faces of those in responsible positions can be seen on the television screen, but there is no possibility of meeting them. Jürgen Habermas *et al.*, *Reflexionen über den Begriff der Beteiligung* (Reflections on the concept of participation: Neuwied, 1961) states that 'the parties . . . are instruments for forming political decisions. These instruments are in the hands not of the people but of those who dominate the party machine.' They confine themselves merely to 'mobilizing the voters temporarily to acts of collective acclamation without in any way altering their lack of political initiative'.

The same observation has been made in the United States, once the home of grass-roots democracy, by R. J. Pranger, *The Eclipse of Citizenship, Power and Participation in Contemporary Politics* (New York, 1968). This state of affairs has been regarded as an inevitable consequence of industrial society by political scientists such as B. N. Luhmann and R. Löwenthal.

R. Gronemeyer, 'Organisierter Alltag, Basisdemokratie oder Eliteherrschaft' (Organized everyday life, basic democracy or domination by an élite) in H. E. Bahr (ed.), *Politisierung des Alltags* (More politics in everyday life: Neuwied, 1972), opposes the view of Luhmann, who considers active participation by every individual in the processes of political decision-making under present-day complex conditions to be a Utopia. Gronemeyer says:

To justify domination by an élite by pointing to the citizen as being incapable seems almost grotesque in the present state of society, where the élite has not even succeeded – to quote but one example – in controlling the deadly chaos of those life-centres of late capitalist society, the cities. The suggestion that the citizen cannot be burdened with participation may therefore be turned round into the assertion that, under the present complexity of circumstances, without participation by the people the avalanche of social problems cannot be stopped.

His aim is 'an increase in institutionalized participation rather than casual participation'. He comes to this conclusion:

> Basic democracy is not some sort of pre-industrial romanticism, outmoded by the functional demands of centralized decision-making in complex organizations. Rather, the everyday problems of production (such as organization of work) and reproduction (organization of planning) demand democratic participation. Without this foundation there are no appropriate solutions. . . . Where powerlessness is overcome, experience creates far-reaching consequences; once the little finger has been grasped the whole hand will follow.

Planning with citizens

The understandable and deeply ingrained opposition to public participation which emerges from some of the sources here cited and is based on the ruling class's fear that it might go too far is probably more deep-seated than the 'obstacles' which allegedly prevent the citizen from becoming a full partner in the democratic process.

Paul Davidoff, one of the leading 'advocate planners', and his students at Hudson College, New York City, told me how quickly those directly affected in planning discussions learn to understand even complicated situations, especially if they have a planner representing their interest at their side as an advocate. He has published several papers in *AIP Journal*, the organ of the American Institute of Planners. In the issue of this same journal for March 1969, H. H. Hyman in his article 'Planning with Citizens' reports on the work of two planning groups with public participation in Boston. In something like 150 public discussions the second group was able to come to a conclusion which satisfied most of the participants. Here the group leader had taken into account the suggestions made by the public; this was in marked contrast to the first group where, from the very beginning, the leader had tried to enforce a view based on that of 'experts'.

The experiences of Paolo Freire, Ivan Illich, Danilo Dolci and, above all, the Chinese planners contradict the legend of citizens unfitted for democracy. The requirements are real sympathy on the part of intellectuals who have different, but not necessarily better, information available, as well as patience, the ability to listen and a good deal of time.

How can the citizen find this time? Has he still the strength, at the end of a working day, to devote himself to his democratic commitments? In the periodical *Bürger im Staat* (The citizen in the state: Bonn) No.3, 1971, Peter C. Dienel, in an article entitled 'Wie konnen Bürger an Planungsprozessen beteiligt werden?' (How can the citizen be involved in the planning process?) makes the suggestion that 'planning cells' should be formed in which the participants are reimbursed for their loss of time:

> For a fully representative planning cell the motivation of voluntary participation is no longer sufficient. The individual should be released by law and reimbursed out of public funds, as is the case with members of parliament, magistrates, national servicemen, and the projected release from work for educational purposes. The costs of four weeks' planning release at the rate of DM 1,500 per person would amount to DM 540,000. In comparison with other costs this does not constitute a financial problem. It will soon be possible to multiply these experiments.

According to Dienel, participants should be chosen as follows:

> The creation of cells which are truly representative and capable of constructive work from the mass of the general public entitled to participate will be made possible by sampling techniques, as used in public opinion polling. It is important that the criteria of selection should be such that they can withstand scrutiny in a court of law. They must be capable of sub-dividing the whole mass of the population, self-evidently fair and incapable of being manipulated. Criteria which would fulfill these conditions would be, for example, date of birth or the initial of the surname. They would allow techniques of selection on which agreement can relatively easily be reached. In taking one birthdate from each decade as the basis, the total population of a *Land* like Hamburg can be reduced to about 360 representatives, who would form twelve 'planning cells' of thirty individuals per cell.

Should a chance bias nevertheless occur in the composition of any one group, this could be evened out thanks to the existence of several groups working on the same problem.

My criticism of Dienel's suggestion is the fact that he envisages that the groups would not formulate their subjects themselves, but would have them assigned by parliament. It is precisely the formation of the citizens' own wishes and initiatives which should be encouraged.

In May 1968 a government committee was set up in Britain under the chairmanship of M. A. Skeffington, MP, to look into and report on the best methods of assuring public participation in the preparatory stages of development plans for each region. The starting point was the consideration that hitherto – as is still the case in most parts of the world – the public, if it has been able to participate in the decision-making process at all, has not been consulted until there is a choice between only a few suggestions previously formulated in long discussions among experts. The report *People and Planning* (HMSO, London, 1969) included a large number of ideas for reform and was generally considered to be a great step forward. It is difficult, however, to put good intentions into practice, as neither officials nor the public have as yet learnt to lose their traditional mistrust of each other. This is borne out by a detailed case study about an experiment in public planning participation in the small town of Millfield, near Sunderland: Norman Dennis, *Public Participation and Planners' Blight* (London, 1972).

Citizens' forums

In 1968 a discussion forum for developmental problems was started in Munich with the aim of becoming 'an agency to bring the ideas of committed citizens to the public notice'. It wants to give all citizens an opportunity to make suggestions for town planning which could then be channelled into the decisions of the administration and the town council. This experiment has been copied in Cologne.

In both towns I asked people where I could find this 'town forum'. Not one of them was able to give me a definite reply, although some of them had 'heard something about it'. The urgent task of revivifying democracy can only be carried through if such forums can be found in the centre of the community and are as well known and accessible as the market-place used to be.

It will be essential to build 'citizens' halls' such as have been suggested by the French artist N. Schoeffer and the Dutchman B. Constant. They should not only contain conference rooms and organize exhibitions and political theatre, but should also contain such technical facilities as to give the public the same 'information power' as the government. A concrete suggestion of what such a meeting place should look like is outlined by H. Hoffmann, 'Plan für ein audiovisuelles Kommunikationszentrum' (A plan for an audiovisual communications centre) in H. Glaser, *Kybernetikon* (Munich, 1972).

An impressive model of a local public information centre was staged under the name 'Capital City Readout' at the 1972 annual meeting of the American Association for the Advancement of Science in Washington D.C. An enlightening summary of this experiment and a discussion of its larger applicability can be found in Thomas V. Vonier and Richard A. Scribner, *Community Information Expositions: issue-oriented displays and popular understanding of social problems* (Washington, 1973).

Information for the public about data banks is contained in H. Sackman, *The Information Utility and Social Choice* (Montval, N. J., 1971).

J. S. Saloma, *System Politics* (Cambridge, Mass., 1968), discusses the possibility that the 'information advantage' of the executive over the legislature might be diminished by allowing Congress to obtain data directly from the government computer (with the excep- of confidential military information).

A 'total information society' which could well turn into a totalitarian one is described in *The Plan for Information Society – A National Goal toward Year* 2000, a faulty translation into English of the Final Report of the Computerization Committee (Tokyo, 1972). In an 'open democracy' this right to information should be available not just to the members of the legislative body, but to every citizen.

At least as important as the installation of the latest information techniques in citizens' forums would be direct personal contact (not lectures!) between the public and politicians, the public and writers, the public and scientists, the public and artists. The isolation of members of the intellectual and power élites should be replaced by frequent contact with the public.

I have called these discussion centres or forums yet to be created 'houses of the future', since it is my hope that in these new places not only present-day worries will be discussed, but also hopes and plans for the world of tomorrow. My idea of these 'workshops for the future', in which the public will develop their wishes and ideas, is outlined in the essay R. Jungk, 'Einige Erfahrungen mit 'Zukunfts-werkstätten' ' (Some experiences with workshops for the future) in *Analysen und Prognosen* (Berlin), January 1973. A more extensive publication is planned.

Information Systems

Technical optimism – social scepticism

The fast progressing development of news transmission and long-distance communications was long regarded as synonymous with an improvement in the possibilities for democracy. This one-sidedly optimistic view has since been superseded by a different and more critical one. It has become clear to many people that the new technological media can be used for the manipulation of the public as well as for serving purposes of information and understanding.

This point is strongly put by Donald Michael of the University of Michigan, 'The Individual: Enriched or Impoverished? Master or Servant?' in a report on a conference under the title *Information Technology – Some Critical Implications for Decision Makers* (The Conference Board, New York, 1971). He maintains that during the next twenty years only a relatively small section of society will, because of its education and social position, actively participate in an 'information-rich' society. An intensive effort to try and include those who continue to reject participation, those who don't care and those who are not qualified to participate – and these still constitute the majority of people – seems to him more important than a further improvement in information technology, which may even, in certain circumstances, be misused for the destruction of freedom and human growth.

The same volume contains an excellent survey of the expected expansion of the 'information environment' by the director of the Center for Integrative Study of the State University of New York at Binghamton: John McHale, 'The Changing Information Environment'. Basing himself on work already in train, he predicts for the period leading up to the year 2000 a strongly converging development of computer and communications technology, with the following possibilities: computer programming by means of the human voice; computers in briefcase size; computers which learn through experience; the feasibility of artificial intelligence; differentiated information through the mass media according to individual requirements; dialogue between the broadcaster and the recipient; and lowering of costs in all branches of information – in the case of computers, down to 1 per cent of previous costs!

Technological optimists and social sceptics discuss their viewpoints in A. F. Westin (ed.), *Information Technology in a Democracy* (Cambridge, Mass., 1971), which contains more than fifty articles surveying the problems of the relationship between the new information systems and the old democratic institutions.

In American specialist journals such as *Datamation* and *Electronics* (New York), technological progress in information technology is viewed positively, even though certain technological shortcomings are conceded, such as overlapping of transmissions.

'Exciting hopes' for a revolution in communications were expressed in a special number of the magazine *International Science and Technology* (New York), April 1968. In this publication J. C. R. Licklider, for instance, developed his ideas for worldwide scientific teamwork on computer communications networks.

Technological aids to democracy

The idea that every citizen should be electronically connected to the centres of political power in order to give him the opportunity of replying originated as far back as the fifties. The concept is set out by V. Zworykin in *The World of 1984* (London, 1964). Anthony Wedgwood Benn, a Cabinet Minister in the British Labour Government, took up the idea and defended it in *New Statesman* (London). But concrete uses of the latest techniques in communications for the processes of democratic decision-making were not tried out until much later, as described in H. Krauch, *Computerdemokratie* (Computer democracy: Düsseldorf, 1972). The author recounts experiments based on his earlier work with W. C. Churchman of the University of California.

The following also deal with this subject:

Amitai Etzioni, 'Minerva: an Electronic Town Hall' in *Policy Sciences*, December 1972;

—— 'An Engineer–Social Science Team at Work' in *Technology Review* (Cambridge, Mass.), January 1975 (a more general discussion of the subject);

V. C. Lamont, University of Illinois, *New Directions for the Teaching*

Computer: Citizen Participation in Community Planning (paper for the World Future Research Conference held at Bucharest in 1972); Chandler Harrison Stevens, Floyd E. Barwig Jr and David S. Haviland, *Feedback – an involvement primer* (National Science Foundation, Washington, January 1974);

Thomas B. Sheridan, *MIT Community Dialogue Project Progress Report* (Cambridge, Mass., July 1974).

The question of how cable television is likely to influence economics, politics and education is dealt with in a factual and informative manner by E. Parker and D. A. Dunn, 'Information Technology, its Social Potential' in *Science* (Washington), 30 June 1972.

Electronics against the city crisis

A most interesting variant of the uses of new electronic means of communication for social purposes was expressed by Peter C. Goldmark, for many years director of the research department of the Columbia Broadcasting System, in a talk given at the first Assembly of the World Future Society in May 1971: Peter C. Goldmark, *New Applications for Communications – Technology for Realizing the New Rural Society*.

The author is of the opinion that more intensive electronic communications would do away with the necessity for the physical concentration of people in a confined area. People could live in the country and still take part in the life of the cities by means of the various information channels. This is how Goldmark envisages it:

We know that most cities today are social networks of communications and if we superimposed on these electronic networks of communications without forethought, this may increase the congestion and complexity of operation. Thus, cities to be built or to be enlarged must be well planned so that the growth is based on a planned network of communications, of transportation, and of utilities.

New communication networks could be divided into internal and external systems. The internal system, which is strictly within the bounds of towns will consist of five basic networks:

The first is the primary network which only exists now in the form of the telephone. In these plans it would be expanded into a full two-way random access network able to accommodate voice, data, and two-way video-phone. This would be the most basic urban 'nerve system' which will be as vital as streets, water, or power. The most basic purpose of this system would be to put anyone in contact with anyone else within the city, no matter how dark the streets are, how heavy the traffic. Through the network's access to data processing, the same system will provide random access between man and machine, or between machines. This network can also be looked upon as providing a pipe into every home, office or library through which one can not only converse, but also transmit and receive written materials, pictures, data, etc. Its most important contribution is its ability of interconnecting any one terminal with any other.

The second network would be in the form of AM–FM radio and television broadcasting, the extent depending on the channels available for the particular community. This could consist of one or more local stations preferably with network affiliations and educational television broadcasting.

The third internal network would be in the form of broadband cables carrying a multitude of television channels into individual homes. This network would include limited address narrow band call-back for purposes of polling or making requests. Such two-way cable television systems are now already under test. The cable network could also carry, if desired, off the air programmes, originating either from local broadcast stations or from satellites. This cable network should be so designed that it has sub-centers at the local neighborhood which, in terms of program material, could cater to its own local audience. As part of this network general informational services would be made available to individual homes. One important example would be the ability to dial up important municipal events, such as meetings of the various town boards, i.e. Education, Finance, Zoning, Board of Representatives, etc. Through the two-way polling ability public opinion could almost be instantly registered on any issue under discussion. Through a system of 'frame freezing' vast amounts of information concerning travel, weather, pollution, shopping, traffic, various municipal and other public services, listing of cultural and entertainment events, etc., could be selected and seen on the home television screen.

The fourth information network superimposed on the town, would be again a broadband cable system, carrying approximately 30 two-way television channels which would interconnect the major public institutions of the city, such as

city hall (center and neighborhood halls)
hospitals and nursing homes
schools and colleges

libraries
police and fire stations
bus, railroad stations and airports
all other town services

The purpose of this would be to provide informational services amongst the vital institutions and key officials of the town for its smooth operation.

Superimposed on these four networks would be a town emergency service. This would include the '911' police and fire emergency system augmented by automatic identification of callers' location and by a system to identify the location of vehicles operated by police, fire, sanitation, ambulance, utilities and other large fleet operations.

Here we have described the internal communication network systems. The external systems would be the following:

1 Incoming broadband cable or microwave circuits which connect the town's businesses, industry and government offices with their operations in other cities or countries. These are essentially dedicated point-to-point links.

2 Long-distance broadband circuits interconnecting the town's switched telephone and video-phone services with the corresponding switched services in other cities.

3 Common carrier broadband and narrow band services such as US Postal Service, Western Union, and others for transmission of messages, printed material, data, etc. between towns and to other countries.

4 Incoming circuits for educational, cultural and recreational pursuits. These would be

(a) radio and television broadcast circuits both for private networks and public broadcasting;

(b) two-way broadband educational television circuits interconnecting a small local campus with the region's central university;

(c) broadband cable circuit as part of a national high-definition closed circuit television network bringing live Broadway, opera, concert and sports productions to theaters especially geared for such performances. The system would employ high-resolution color television of at least 1,000 lines with cameras and projectors especially designed for live pick-up and large screen projection. The most suitable national distribution method for such signals may be through a synchronous satellite broadcasting several of these 'high-definition TV signals and received by local high-gain fixed antennas.

Counter-information from the underground

Networks of information will soon be spanning the earth. In spite of considerable difficulties, efforts to introduce a worldwide science information system are gradually succeeding. A brief report on this is contained in UNISIST, *Synopsis of the Feasibility Study on a World Science Information System* (UNESCO, Paris, 1971).

Moreover various countries, including Britain, Japan, France and West Germany, are engaged in preparations for the creation of national information networks. There could, however, be political dangers inherent in a system of this kind covering every home, especially when linked to central data banks capable of storing records about every citizen. This is convincingly argued by Nigel Calder, *Technopolis* (London, 1969), and A. E. Miller, *The Assault on Privacy* (Ann Arbor, Michigan, 1971).

Professor Robert M. Fano of MIT comments on the necessity for democratization of data equipment, saying that the maintenance of a sensible balance of power in society, and hence the maintenance of personal freedom, will in future depend heavily on whether the services of powerful computer systems are made generally available, practically and economically, like electricity or the telephone today.

If, however, not everyone has access to the new information systems, private networks will be started. Beginnings are already being made.

Such a development of information networks and data banks available to radical groups in society would at present seem to be a somewhat adventurous idea. In fact it would merely correspond to the setting up last century of a critical opposition press, which has long been accepted as a matter of course.

The attempts to develop 'alternative television' have been forcefully depicted in a French film reportage by C. de Givray and P. Schaeffer, 'Vingt millions de caméras citoyens?' (Twenty million citizen cameras), which was shown at the Prix Futura television contest in Berlin. The printed script is available.

Thoughts about democratizing electronic mass media were already expressed by Bertolt Brecht, and were further developed by Hans Magnus Enzensberger, 'Baukasten zu einer Theorie der Medien' (Tools for a media theory) in *Kursbuch* (Berlin) No. 20, 1970. The pioneering experiment 'Kybernetikon', run by the town of

Nuremberg in 1972 under the direction of H. Glaser and K. H. Stahl, concerned itself with these theories with the participation of the public. Information about ideas and experiments in this field can be found as follows:

Britain
B. Jardine and M. Hickie, 'The North Kensington Project' in *Architectural Design* (London), November 1972;
B. Groombridge, *Television and the People – A Programme for Democratic Participation* (London, 1972);
J. Hopkins, *Video in Community Development* (London, 1972).

France
P. Lewis, 'Community TV – a New Hope' in *New Society* (London), 9 March 1972;
P. Schaeffer, *Pouvoir et communication* (Power and communication: Paris, 1972.

Canada
Challenge for Change (Toronto), 1971– .

USA
N. Johnson, *How to Talk Back to your TV Set* (Boston, 1970);
M. Hinshaw, *Wiring Megalopolis* (Annenberg Fund, 1972);
M. Shamberg, *Guerrilla Television* (New York, 1971);
M. Price, *Cable TV: a Guide for Citizen Action* (Philadelphia, 1972);
Radical Software (New York): also with detailed reports on Canadian experiments;
Prime Time (San Francisco, Box 630).

Those interested in an exchange of videotapes can apply to the Cultural Data Bank, Raindance Video Service, PO Box 5423, Cooper Station, New York 1003.

Information about opposition to cable TV in the United States is contained in:
S. P. Suchermann, 'Cable TV: the Endangered Revolution' in *Columbia Journalism Review*, May 1971;
T. Meehan, 'Coming up on Channel C: You!' in *Saturday Review* (New York), 9 September 1972;
Columbia University, *Cable TV – the end of a dream* (New York, 1974).

The counter-culture network

The change in the mental climate, the infiltration into society of alternative thought and the resulting change in attitudes have become increasingly noticeable since the middle sixties. How does this process come about? By what means and through what network of communication? Several American authors have tried to find explanations. The best of these is D. Schon, *Beyond the Stable State* (London, 1971). The author tries to explain that centrally controlled networks, through which orders and instructions are passed from the centre to outer control points, are no longer able to guide the enlarged and far more varied economic and social systems of the late twentieth century.

The 'movement' of American youth against the economic and technological system of their fathers is seen by Schon as a model for the new way in which innovatory ideas are spreading. There is no longer one directing centre, but many. They change frequently, as do their leadership and composition, and their ideas are not programmed but in a constant process of development. Thus a living system arises, able to survive because of its flexibility, its constant readiness to learn and to adapt. It makes partial use of the technological infrastructure of information and communications systems (television, radio, telephone, tape and video recorders, records, postal communications) but is at the same time based on personal relationships, the spoken word from man to man.

One of the leading figures of this movement has described this process from his own experience and daily struggle: Michael Rossman, *On Learning and Social Change* (New York, 1972), with special reference to the chapter 'A Communications Network for Change in Higher Education'. The same author has developed his ideas about spreading this new way of thinking in a convincing and concise article in the *Saturday Review* (New York), 19 August 1972, under the title 'How we Learn today in America'.

In order to make this network of widely scattered, frequently replaced 'growth centres' more easily accessible, a number of address lists have appeared which, through their comments, give an idea of the variety of the decentralized learning process as analysed by Schon. An example is *The People's Yellow Pages* – named after the yellow pages in telephone directories which list commercial undertakings – a catalogue appearing in the Boston and Cambridge region and published by 'Vocations for Social Change', 351 Broadway, Cambridge, Mass. It lists the addresses of individuals and groups who offer their services 'without intention of exploitation': people with some knowledge of music, carpentry, art or medicine, groups concerned with non-violent action, environmentalists, free schools.

The best survey, to my mind, is contained in *Source Catalog – Communications* (PO Box 21066, Washington, D.C. 20009). A team of the Education Liberation Front (ELF) who travelled by bus from one university to another made up this list when they had collected too much information to publicize from a bus. Trained librarians of the American counter-culture composed an impressive mosaic from hundreds of brief self-portraits of groups and institutions of the counter-culture. Their motivation: 'We need information about our efforts, our means, projects, abilities and dreams, in order to know the infrastructures of help which are necessary for us to free this country and ourselves.' They list news agencies of the underground press, pamphlets and books about the role of the mass media, groups of artists and craftsmen, and so on.

Invisible colleges

To conclude my notes on alternative information networks, I should like to mention that my attention was drawn to *Source*, as well as to many other items, by a man whose serious hobby it has been for many years to run his own worldwide information network. In the mid-sixties John Dixon, head of the department for educational projects at Xerox in Washington – where he had the opportunity to make use of the firm's copying machines without charge – began to send copies of newspaper and magazine articles on themes concerning the future to public personalities and friends. He thereby substantially contributed to the creation of international futurological research.

For in this way one of the 'invisible colleges' was formed which the American historian Derek de Solla Price has described as decisive for the progress of science. Through letters, photocopies of manuscripts, telephone conversations and conferences, 'in-groups' are formed within nearly all branches of research which pass on information about ideas, plans and work projects more rapidly than the specialized journals ever could. In this way the most advanced researchers in all fields can find out what is going on without drowning in a flood of printed matter. A most interesting book about this phenomenon is Diana Crane, *Invisible Colleges: Diffusion of Knowledge in Scientific Communities* (Chicago, 1972).

Man's Future

The beginnings of humanistic futurology

The crisis resulting from largely uncontrolled scientific and technological developments was the cause of attempts to start futurological research. In this context man proved a factor hard, perhaps impossible, to comprehend. In my article 'Human Futures' in *Futures* (Guildford, Surrey), September 1968, I attempted to criticize the one-sidedness and incompleteness of a futurology which does not pay sufficient attention to the human factor.

Teilhard de Chardin and Julian Huxley have done pioneering work in this field. The work of Pierre Bertaux, *Mutation der Menschheit* (Mutation of humanity: new ed. Bern, 1967), takes an important step in that direction. The author points out the increasing importance of the group as against the individual.

But systematic work on a more human treatment of prognoses for the future, especially in the cultural field, only began with the symposium 'Cultural Futurology' which was held in November 1970 as part of a conference of the American Anthropological Association. It had been prepared by an earlier exchange of detailed scientific papers among the participants. These beginnings have been outlined by M. Maruyama and J. A. Dator (eds.), *Human Futuristics* (University of Hawaii, 1971). An important point seemed to me to be the warning by Maruyama not to approach future man and his culture with the same expectations and standards as are applied to existing attempts at obtaining knowledge about the future.

An important new approach tried by Johan Galtung is described in his article 'Towards New Indicators of Development' in *Futures*, June 1976. He stresses 'human needs' as an important indicator of development, and emphasizes, besides material requirements, such needs as:

◇ equality and social justice;

◇ level of autonomy or self-reliance;
◇ participation of all.

Attempts to develop humanistic futuristics further, and to create a counterweight to mainly economic and technological concepts of the future, have been made by:

The Teilhard de Chardin Centre for the Study of Man, London;
Mankind 2000, Brussels (Secretary-General J. Wellesley-Wesley);
Center for Integrative Studies, State University of New York, Binghamton (J. McHale and Magda Cordell Hale);
Social Science Research Institute, Honolulu;
'Fondation Royaumont pour le progrès des sciences de l'homme' (Royaumont Foundation for the progress of the sciences of man), Paris (Secretary-General M. Piatelli-Palmarini);
The Educational Policy Research Center of the Stanford Research Institute, Palo Alto, California, under the direction of W. Harman.

Changing values

Of great importance in human futuristics is the study of value changes. The foundations for this were laid by Wolfgang Köhler, who emigrated from Berlin to the USA, in *The Place of Value in a World of Fact* (new ed., New York, 1968).

An essay by I. Taviss, 'Futurology and the Problem of Values' in *International Social Science Journal* (UNESCO, Paris) No. 4, 1969, gives an excellent introduction to the relationship between values and futurology. So does the bibliography I. Taviss, *Technology and Values* (Cambridge, Mass., 1969), compiled in the context of the Harvard University Program on Technology and Society at the University of Pittsburgh.

A sizeable volume of collected papers giving information about attempts to forecast future changes in attitudes to values is K. Baier and N. Rescher (eds.), *Values and the Future* (New York, 1969). A detailed and stimulating introduction to this volume is supplied by Alvin Toffler under the title 'Value Impact Forecaster – a Profession for the Future'.

Also of interest are:

A. J. Wiener, *Changing Values and Social Goals* (Hudson Institute, Croton, 1970);
A. Mitchell and M. K. Baird, *American Values* (Stanford Research Institute, Palo Alto, California, 1969);
Daniel Yankelovich, *The New Morality: a profile of American youth in the 70s* (New York, 1974).

The science of man

In the first instance, I should like to mention the researches of Edgar Morin (already referred to in the text), which cover a large number and variety of the branches of the science of man in a new comprehensive view. This is especially the case in his latest work, *Le Paradigme perdu: la nature humaine* (The lost paradigm: human nature: Paris, 1973), which contains a very detailed bibliography. It could provide a basis for an interdisciplinary science of man. The book is prefaced by the following 'timetable':

Universe	7,000 million years
Earth	5,000 million years
Life	2,000 million years
Vertebrates	600 million years
Reptiles	300 million years
Mammals	200 million years
Anthropoids	10 million years
Hominides	4 million years
Homo sapiens	100,000 to 50,000 years
City, state	10,000 years
Philosophy	2,500 years
Science of man	0 years

The address of Morin's Institute is 23 bis rue de l'Assomption, Paris, and one of its latest publications is: Edgar Morin and Massimo Piatelli-Palmerini, *L'Unité de l'homme* (The unity of man: Paris, 1974).

Also of importance are Jonas Salk, *Man Unfolding* (New York, 1972), and W. W. Harman and the staff of the Centre for the Study of Social Policy, *Changing Images of Man* (Menlo Park, California, 1974).

Index

Figures in italics refer to entries in Part Two, the Tool-Kit.